Key Concepts in Mathematics

Key Concepts in Mathematics

Alberto D. Yazon

ARCLER
PRESS

www.arclerpress.com

Key Concepts in Mathematics

Alberto D. Yazon

Arcler Press
224 Shoreacres Road
Burlington, ON L7L 2H2
Canada
www.arclerpress.com
Email: orders@arclereducation.com

ISBN: 978-1-77469-191-5 (Hardcover)

ABOUT THE AUTHOR

Dr. Alberto Dolor Yazon is an Associate Professor and Chairperson of Curriculum and Instruction Development and Quality Assurance at the Laguna State Polytechnic University, Philippines. He obtained his Doctor of Philosophy in Mathematics Education at Philippine Normal University, Manila. He is a Technical Committee and Advisory Board Member of EURASIA Research since September, 2019. He authored the published books Learning Guide in Methods of Research, Assessment in Student Learning, Fundamentals of Advanced Mathematics, and Introduction to Mathematical Literacy. He had published 13 research papers, three (3) of which are indexed in Scopus Journal. He is a resource speaker for different topics related to Mathematics, Assessment of Learning, Statistics, and Research. He seats as a panel of examiner and adviser in master's thesis and doctoral dissertation among graduate students. He is also an Accreditor in the Accrediting Agency of Chartered Colleges and Universities in the Philippines, Inc. (AACCUP) and Internal Quality Auditor in International Organization for Standardization (ISO).

TABLE OF CONTENTS

LIST OF FIGURES

LIST OF ABBREVIATIONS

CP	Cost Price
DDMF	Dynamic Dictionary of Mathematical Functions
GCD	Greatest Common Divisor
HCF	Highest Common Factor
IQR	Interquartile Range
MSD	Mean Squared Deviation
MSE	Mean Squared Error
RMSE	Root-Mean-Square Error
RMSD	Root-Mean-Square Deviation
ROC	Rate of Change
SI	International System of Units

PREFACE

Mathematics is a science developed around theories and concepts of the logic of form, quantity, and arrangement. Mathematics applies to everyday life, to everything the individual does. It is the basic element of everything in daily life. In addition to sports, it also includes mobile devices, ancient or modern architecture, art, money, and engineering.

Since the earliest period of recorded history, discoveries in mathematics have been at the forefront of all advanced societies and have been applied even in the oldest cultures. The requirements of mathematics depend on the needs of society. The complexity of mathematical needs is more related to the complexity of society. Primitive tribes only need the ability to count, but they also rely on the mathematics to calculate the position of the sun and the physics of hunting.

Mathematics is the science of structure, order, and relations. It has evolved from the main practice of counting, measuring, and defining the shape of objects. It handles logical reasoning while processing quantitative calculus, and its evolution has incorporated more and more idealization and abstraction of its subject matter. Throughout the 17th century, mathematics was not only an indispensable supplement to physical science, but also an indispensable supplement to technology. Today it plays the same role in the quantitative dimension of life sciences.

In various cultures, under the stimulation of practical activities such as business and agriculture, the development of mathematics has far exceeded basic counting. This development is a huge aspect in a society that is complex enough to sustain these activities, free time for examinations, and opportunities to build on the achievements of ancient mathematicians.

All mathematical systems (such as Euclidean geometry) are a combination of axioms and theorem sets, which can be logically derived and identified from the axioms. The study of mathematical logic and philosophical foundations seeks questions and answers about whether the axioms of the proportional system ensure its integrity and stability.

Due to the exponential development of science, most of the mathematics has been developed since the 15th century AD, and the fact that has been stated in history is that from the 15th century to the end of the 20th century, the new expansion of mathematics was mainly concentrated in Europe and North America. For these special reasons, all dimensions and aspects of this book are dedicated to the development of Europe starting in 1500.

However, this does not mean that growth and evolution elsewhere are trivial. In fact, to understand the history of mathematics in Europe, you must at least understand its history in early Mesopotamia and Egypt, as well as ancient Greece and the Islamic civilization that dates back to the 9th to 15th centuries.

India's role in the development of modern mathematics is due to the significant influence of India's achievements in Islamic mathematics during its constructive period. On the other hand, South Asian mathematics focuses on the earlier or ancient history of mathematics in the Indian subcontinent and the evolution of contemporary place value decimal numbering systems.

This book will introduce readers to the field of mathematics and its basic aspects. It is designed for readers or students who are inexperienced in mathematics and its various dimensions, and it covers a variety of basic topics. After reading this book, readers will understand the basic knowledge of mathematics and the key concepts involved.

A

Absolute Value

The absolute value of a number is given by distance measured along the number line from the origin 0 to the number. The absolute value of 3 at point B is the distance of point B from 0, which is 3. The absolute value of –2 at point A is the distance between point A and zero, which is 2.

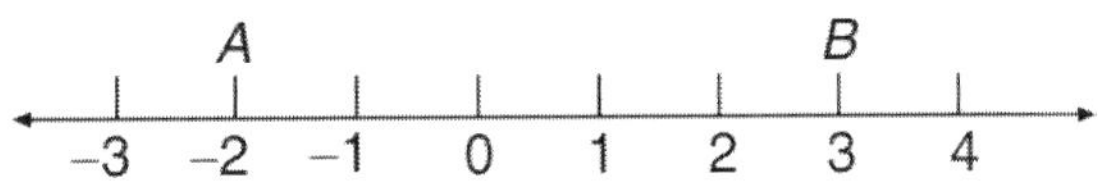

The absolute value of a number is represented by two vertical parallel lines placed around the number. The absolute value of 3 at point B is given as

|point B − point 0| = |3 − 0|

= |3|

= 3

If the subtraction was done in the reverse direction, the solution would still be 3, but the working would be

Absolute value of 3 = |point 0 − point B|

= |0 − 3|

= |−3|

= 3 as before

Likewise for the point A:

Absolute value of −2 = |point A − point 0|

= |−2 − 0|

= |−2|

= 2

The absolute value of the line segment AB is given as:

Absolute value of AB = |point A − point B| or |point B − point A|

= |−2 − 3|

= |−5|

= 5

When an absolute value is used in subtraction, the two numbers on the number line can be subtracted in either direction; the length of the line segment that connects them yields the same response.

Abstract

The term "abstract" refers to the separation of real problems from the real world. Mathematics is an abstract subject, practiced through the use of symbols, although these symbols can represent objects in the real world. Abstract mathematics can be used to solve real-world problems, and the symbols used can be used to solve practical problems. The real problem in the following example is determining the width of the lawn. The symbol for this width in the summary is x. The value of x is determined using abstract mathematics, so the width of the grass can be calculated and a solution can be found.

Abstract Algebra

In modern mathematics, algebraic structure is considered a set that defines operations, and algebraic concepts generally related to real number systems are extended to other more general systems, such as groups, rings, fields, modules, and vector spaces. Abstract algebra is a broad field of mathematics that involves algebraic structures such as groups, rings, vector spaces, and algebra. Generally speaking, abstract algebra is the study of what happens when certain properties are abstracted from a number system; for example, changing the definition of basic arithmetic operations will produce a structure called a ring, as long as the operations are consistent.

For example, a 12-hour clock is an example of such an object, in which the arithmetic operations have been redefined to use modular arithmetic (modulo 12). An additional level of abstraction that considers a single operation allows the clock to be understood as a group. In either case, abstraction is useful because many properties can be understood without considering the specific structure in question. This is especially important when considering the relationship between structures; the concept of group isomorphism is an example.

Acceleration

When a car increases its speed, such as when changing lanes on a highway, the rate at which its speed changes over time is the acceleration of the car. The SI unit of acceleration is meters per second, or m/s^2 for short. Centimeters per second, written cm s^2 or cm/s^2, is another unit of acceleration. When an object falls on the earth, it will experience a gravitational acceleration of about 10 m s^2, and this acceleration is constant at different points on the earth's surface. This means that the speed of the stone falling in the air increases by 10 ms^1 for every second of falling.

The acceleration is represented by the slope of the speed-time graph. When a stone is thrown upwards, its speed will decrease, and the negative acceleration is called deceleration. The speed-time graph is a curve when the acceleration is not constant but changes. For example, this is the case when a motorcyclist accelerates from a standstill.

Accuracy

Due to the obvious limitations of the measuring equipment, the measured dimensions, such as the height of the structure, cannot be accurate. The measured value is estimated and presented with a certain degree of accuracy in the form of decimal places (dp), significant figures (sf) or the nearest integer. When making measurements or calculating with measured quantities, the accuracy should be mentioned in the answer. This method of providing an estimated answer is called rounding.

Acre

The acre is an imperial unit of area used to measure the size of a piece of land. One acre is equal to 4840 square yards (43,560 square feet). Traditionally, an acre was described as the quantity of ground that could be ploughed in one day by a pair of oxen. One acre equals approximately 0.40 hectare, or one hectare equals approximately 2.47 acres. In practice, 1 hectare equals around 21 2 acres, and 5 acres equals approximately 2 hectares.

Acute Angle

This topic also covers right angles, straight angles, obtuse angles, and reflex angles. To explain acute angles, it is essential to first define a right angle. A right angle is defined as an angle of 90 degrees, which is written as 90°. A right angle represents a quarter turn. Acute angles are angles that are less than one right angle, or less than 90 degrees. The lid in a case will fall back to top of the box when the lid is opened by an acute angle and subsequently dropped.

Figure 1. Acute angle.

Source: Image by Noun Project.

Addend

Addends refer to the numbers or quantities which are added together. The resulting answer is denoted as the sum:

Addend + addend = sum

A problem may involve more than two addends. For instance, in order to receive the sum of 3 + 9 + 6, the pair of numbers 3 and 9 is added to obtain 12, and the 12 is added to 6 for obtaining 18. Or, through mental addition, one could build the ability to make them into pairs that sum 10, as adding numbers to 10 is easier.

Adjacent Angles

Adjacent angles are angles that are side by side. When two adjacent angles form a straight line they add up to 180° is said to be adjacent angle.

Algebra

The abstract study of the properties of numbers with letters to represent the numbers is known as algebra; these letters are referred to as variables. Variables represent unknown values, and arithmetic operations are employed to try to determine their value. This entry describes how expressions can be simplified or rewritten by adding, subtracting, multiplying, and dividing terms.

Algebraic Fractions

Like arithmetic fractions, algebraic fractions can be cancelled, added, subtracted, multiplied, and divided. The entry Canceling explains how to cancel fractions. It is useful to remember that if two fractions are equivalent, they can be represented in several ways, as seen here.

1/5 y is the output of multiplying the variable y by the fraction 1/5. When 5 is divided by y, the result is expressed as y/ 5.

As multiplication by 1/5 produces the same result as dividing by 5, 1/5 y and y/5 are equal. Likewise, 2/ 3 x and 2x/3 are equal.

Alternate Angles

Figure 2 depicts two parallel lines marked by arrows. A transversal is a line that passes through them. A pair of equal-sized angles, such as a and b, that occur on opposite sides of the transversal and lie between the parallel lines are referred to as alternate angles.

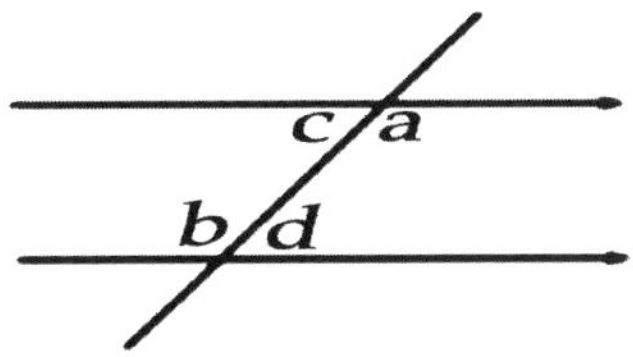

Figure 2. Two parallel lines marked by arrows.

Source: Image by Wikimedia commons.

A reversed Z-like shape is formed by alternate angles, as depicted in Figure 3.

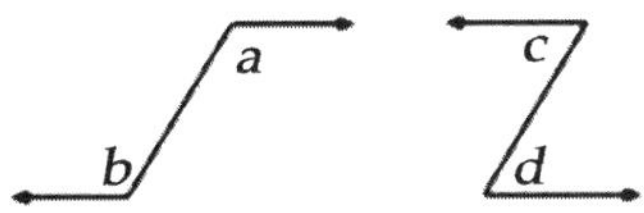

Figure 3. Reversed Z-like shape is formed by alternate angles.

Source: Image by Wikimedia commons.

Angle a = angle b

The pair of alternate angles shown in figure c forms a letter z:

Angle c = angle d

Altitude

This word can be used in two ways. The altitude of an object refers to the distance of the object beyond the earth's surface, also known as its vertical height. Altitude has a somewhat different meaning in geometry – it refers to the altitude of a polygon or a polyhedron. The term altitude is defined in this context under the entry Base (geometry).

Amicable Numbers

Pairs of numbers for which the sum of the divisors of one number equals the other number, e.g., 220 and 284, 1184 and 1210.

The pair of numbers 220 and 284 have the curious property that each "contains" the other. In what way? In the sense that the sum of the proper positive divisors of each, sum to the other.

Such pairs of numbers are called amicable numbers (amicable means friendly--but there is a different set of number actually called friendly number).

Amicable numbers have a long history in magic and astrology, making love potions and talismans. As an example, some ancient Jewish commentators thought that Jacob gave his brother 220 sheep (200 female and 20 male) when he was afraid his brother was going to kill him (Genesis 32:14). The philosopher Iamblichus of Chalcis (ca. 250–330 A.D.) writes that the Pythagoreans knew of these numbers:

They call certain numbers amicable numbers, adopting virtues and social qualities to numbers, such as 284 and 220; for the parts of each have the power to generate the other.

Pythagoras is reported to have said that a friend is "one who is the other I, such as are 220 and 284." Now amicable numbers are most often (and most properly!) relegated to the exercise sections of elementary number theory texts.

There is no formula or method known to list all of the amicable numbers, but formulas for certain special types have been discovered throughout the years. Thabit ibn Kurrah (ca. A.D. 850) noted that

if $n > 1$ and each of $p = 3.2^{n-1}-1$, $q = 3.2^{n}-1$, and $r = 9.2^{2n-1}-1$ are prime, then $2^{n}pq$ and $2^{n}r$ are amicable numbers.

It was centuries before this formula produced the second and third pair of amicable numbers! Fermat announced the pair 17,296 and 18,416 (n=4) in a letter to Mersenne in 1636. Descartes wrote to Mersenne in 1638 with the pair 9,363,584 and 9,437,056 (n=7). Euler then topped them both by adding a list of sixty-four new amicable pairs, however he made two errors. In 1909 one of his pairs was found to be not amicable, and in 1914 the same fate took a second pair. In 1866 a sixteen-year-old boy, Nicolo Paganini, discovered the pair (1184,1210) which was previously unknown.

Now extensive computer searches have found all such numbers with 10 or fewer digits and numerous larger examples, for a total of over 7500 amicable pairs. It is unknown if there are infinitely many pairs of amicable numbers. It is also unknown if there is a relatively prime pair of amicable numbers. If there is such a pair, they must be more than twenty-five digits long, and their product must be divisible by at least 22 distinct primes.

Amplitude

Amplitude is a characteristic of periodic curves such as the sine and cosine curves. The amplitude of the sine curve is the largest distance of a point on the curve from the x-axis, and it is represented in the figure by a. The amplitude of the sine curve y = sin x is a = 1. The amplitude of the curve y = 2 cos x is a = 2.

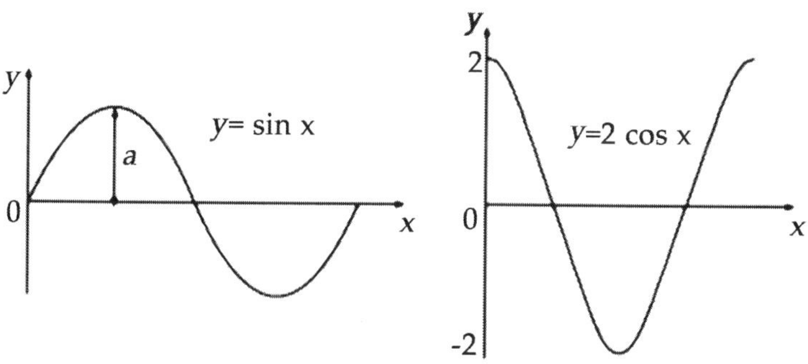

Figure 4. Amplitude.

Source: Image by Wikipedia.

Analytic (Cartesian) Geometry

The study of geometry using a coordinate system and the principles of algebra and analysis, thus defining geometrical shapes in a numerical way and extracting numerical information from that representation.

Analytic geometry, also called coordinate geometry, mathematical subject in which algebraic symbolism and methods are used to represent and solve problems in geometry. The importance of analytic geometry is that it establishes a correspondence between geometric curves and algebraic equations. This correspondence makes it possible to reformulate problems in geometry as equivalent problems in algebra, and vice versa; the methods of either subject can then be used to solve problems in the other. For example, computers create animations for display in games and films by manipulating algebraic equations.

Analysis (Mathematical Analysis)

Grounded in the rigorous formulation of calculus, analysis is the branch of pure mathematics concerned with the notion of a limit (whether of a sequence or of a function).

Analysis, a branch of mathematics that deals with continuous change and with certain general types of processes that have emerged from the study of continuous change, such as limits, differentiation, and integration. Since the discovery of the differential and integral calculus by Isaac Newton and Gottfried Wilhelm Leibniz at the end of the 17th century, analysis has grown into an enormous and central field of mathematical research, with applications throughout the sciences and in areas such as finance, economics, and sociology.

The historical origins of analysis can be found in attempts to calculate spatial quantities such as the length of a curved line or the area enclosed by a curve. These problems can be stated purely as questions of mathematical technique, but they have a far wider importance because they possess a broad variety of interpretations in the physical world. The area inside a curve, for instance, is of direct interest in land measurement: how many acres does an irregularly shaped plot of land contain? But the same technique also determines the mass of a uniform sheet of material bounded by some chosen curve or the quantity of paint needed to cover an irregularly shaped surface. Less obviously, these techniques can be used to find the total distance traveled by a vehicle moving at varying speeds, the depth at which a ship will float when placed in the sea, or the total fuel consumption of a rocket.

Similarly, the mathematical technique for finding a tangent line to a curve at a given point can also be used to calculate the steepness of a curved hill or the angle through which a moving boat must turn to avoid a collision. Less directly, it is related to the extremely important question of the calculation of instantaneous velocity or other instantaneous rates of change, such as the cooling of a warm object in a cold room or the propagation of a disease organism through a human population.

Angle

Geometry is a branch of mathematics that deals with the study of shapes and their measurements. It also focuses on the relative configuration of the shapes and their spatial properties. We know that geometry is classified into 2D Geometry and 3D Geometry. Before dividing that, all the geometrical shapes are formed by points, lines, rays and plane surface. When the two lines or the rays converge at a common point, the measurement between the two lines is called an "Angle."

In Plane Geometry, a figure which is formed by two rays or lines that shares a common endpoint is called an angle. The word "angle" is derived from the Latin word "angulus," which means "corner." The two rays are called the sides of an angle, and the common endpoint is called the vertex. The angle that lies in the plane does not have to be in the Euclidean space. In case if the angles are formed by the intersection of two planes in the Euclidean or the other space, the angles are considered dihedral angles. The angle is represented using the symbol "∠." The angle measurement between the two rays can be denoted using the Greek letter θ, α, β etc. If the angles are measured from a line, we can find two different types of angles, such as a positive angle and a negative angle.

Positive Angle: If the angle goes in counterclockwise, then it is called a positive angle.

Negative Angle: If the angle goes clockwise direction, then it is called a negative angle.

The angles are classified under the following types:

1. Acute Angle – an angle measure less than 90 degrees.
2. Right Angle – an angle is exactly at 90 degrees.
3. Obtuse Angle – an angle whose measure is greater than 90 degrees and less than 180 degrees.
4. Straight Angle – an angle which is exactly at 180 degrees.
5. Reflex Angle – an angle whose measure is greater than 180 degrees and less than 360 degrees.
6. Full Angle – an angle whose measure is exactly at 360 degrees.

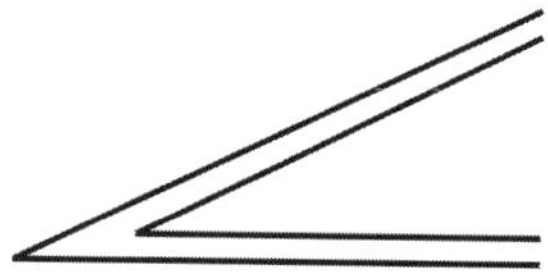

Figure 5. Angle.

Source: Image by Pixabay.

Annulus

An annulus is a shape made out of two circles. It is a plane figure formed by two concentric circles. The region covered between two concentric circles is called annulus. It has a ring shape and has many applications in Mathematics. Some of the real-life examples are finger-ring, dough-nut, etc. The area of the annulus is determined if we know the area of circles (both inner and outer). The formula to find annulus of the circle is given by:

$$\mathbf{A = \pi(R^2 - r^2)}$$

where 'R' is the radius of outer circle and 'r' is the radius of inner circle.

The circle is a fundamental concept not only in Math's but also in many fields. By its definition, we know, a circle is a plane figure that is made up of the points situated at the same distance from a particular point.

It shows a complete circle with some radius. Now if the same circle is surrounded by another circle with some space in between them and radius bigger than this circle, the region formed in between the two circles is basically the annulus.

The word "annulus" (plural – annuli) is derived from the Latin word, which means "little ring." An annulus is called the area between two concentric circles (circles whose center coincide) lying in the same plane.

It is the region bounded between two circles which share the same center. This shape resembles a flat ring. It can also be considered as a circular disk having a circular hole in the middle.

Area of Annulus

The area of the annulus can be calculated by finding the area of the outer circle and the inner circle. Then we have to subtract the areas of both the circles to get the result. Let us consider Figure 6.

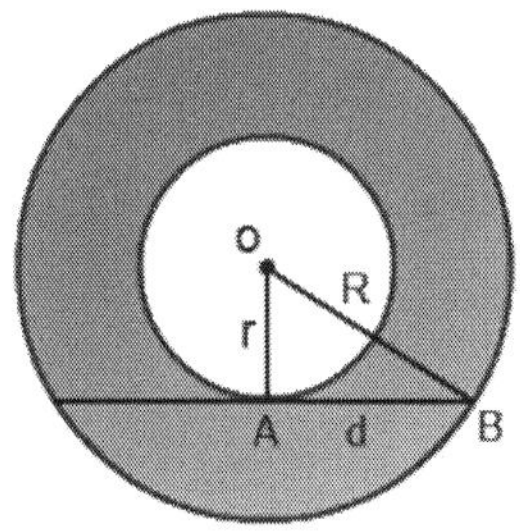

Figure 6. Area of annulus.

Source: Image by Flickr.

In the above figure, two circles are having common center O. Let the radius of outer circle be "R" and the radius of inner circle be "r." The shaded portion indicates an annulus. To find the area of this annulus, we are required to find the areas of the circles.

Therefore,

Area of Outer Circle = πR^2

Area of Inner Circle = πr^2

Area of Annulus = Area of Outer Circle – Area of Inner Circle

Hence,

Area of Annulus = $\pi (R^2 - r^2)$

Area of Annulus = $\pi (R+r) (R-r)$

Angle Between a Line and a Plane

This is a subtopic under three-dimensional trigonometry. The angle between a line and a plane is obtained by projecting the line onto the plane and then calculating the angle between the projected line and the original line. This is the angle formed by the line and the plane. Assume a straight nail ON is hammered at an angle into a piece of wood, such that the nail is not straight (refer to figure a). Figure a shows the projection of the nail ON onto the plane of the wood. The shadow created by the nail ON when parallel rays of light shine at right angles to the plane can be thought of as ON's projection onto the plane. The angle NOW is defined as the angle formed by the line ON and the plane of the wood.

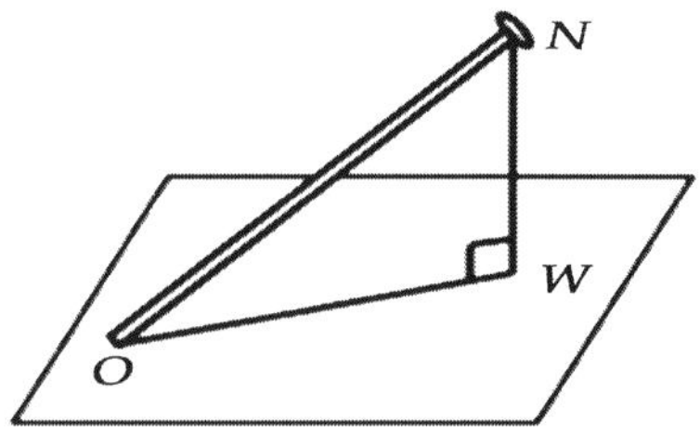

Figure 7. Angle between a line and a plane.

Source: Image by Wikimedia commons.

Angle Between Two Planes

This is again a subtopic under three-dimensional trigonometry. Assume that two planes that are inclined to each other cross in the straight line XY. The two planes can be conceived of as held together, like a trapdoor that opens. If the sloping plane drops down onto the other plane, the angle at which it turns is equal to the angle between the two planes. With regard to the two planes, this angle is also known as the dihedral angle.

Angle Bisector

A line that partitions an angle into two equal parts. The angle bisector is built with a ruler and compass.

Angle In a Semicircle

A triangle drawn inside a semicircle is the focus of this geometry theorem. This theorem is a subset of another presented theorem under the entry Angle at the Center and Circumference of a Circle.

Angle of Depression

Assume a surveyor is standing at point A on top of a wall, facing horizontally out to sea. The angle x via which she lowers her sight to look at a buoy B out at sea represents the Buoy's depression from her point A.

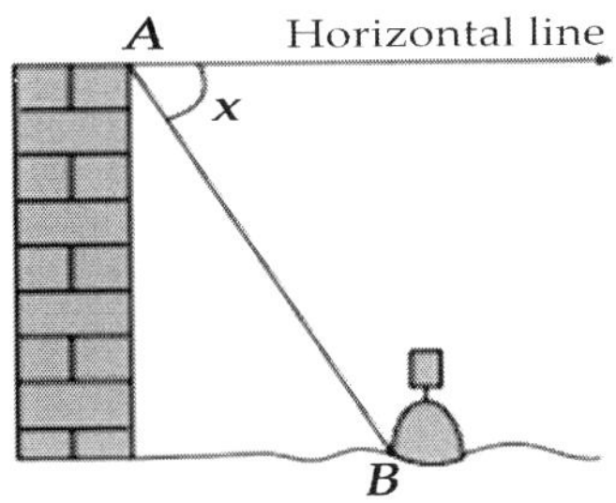

Figure 8. Angle of depression.

Source: Image by Wikimedia commons.

Angle of Elevation

Assume Darren has gone to the top of his house and is looking out into the distance horizontally. The angle y at which he raises his eyes to look up at the top of a flagpole is known as the angle of elevation of the flagpole from his location A.

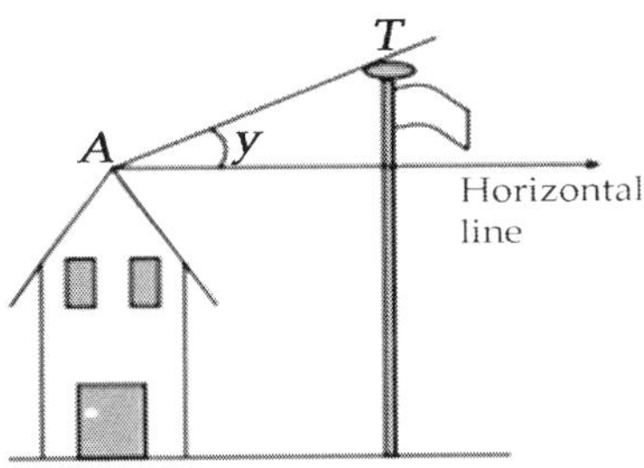

Figure 9. Angle of elevation.

Source: Image by Flickr.

Angle of Inclination

This is the angle formed by a particular line with another line or with a plane. Assume Helen is climbing down a building's wall and her rope forms a 33-degree angle with the wall. The rope's angle of inclination to the wall is 33°. In other words, the rope is angled at 33 degrees to the wall. The term "angle of inclination" can also be used to represent the angle formed by two parallel planes.

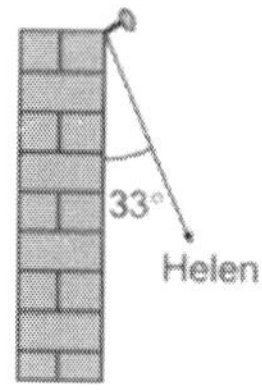

Figure 10. Angle of inclination.

Source: Image by Wikimedia commons.

Angle Sum of a Triangle

A geometry theorem that asserts that the sum of a triangle's three angles is 180°. This can also be stated in another way: A triangle's three angles are supplementary. The following experiment can be used to illustrate this geometry theorem: Draw a triangle on a piece of paper and carefully cut it out using scissors. Further cut each of the three angles A, B, and C and rearrange them as indicated in figure a to see that they create a straight line with an angle of 180°.

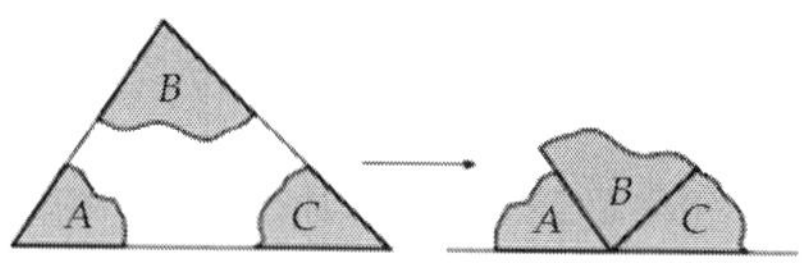

Figure 11. Angle sum of a triangle.

Source: Image by Wikipedia.

Angles At a Point

Angles at a point are formed by two or more angles that meet at a point and combine to form a full turn. The three angles 100°, 95°, and x in figure a represent angles at a point. When two or more angles converge at a point and combine to form a full turn, the sum of the angles is 360°, because a full turn is of 360°. Angles whose sum is 360° are known as conjugate angles. As seen in figure a,

$x + 100° + 95° = 360°$ Sum of angles at a point = 360°

$x = 165°$

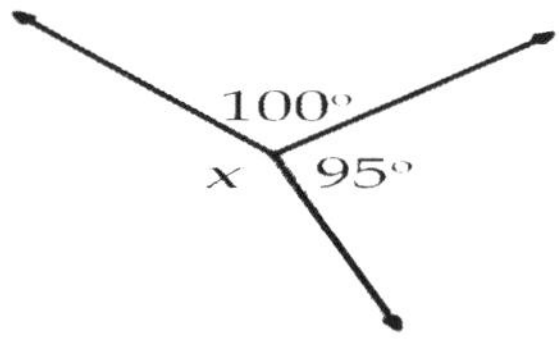

Figure 12. Angles at a point.

Source: Image by Flickr.

Angles On the Same Arc

This entry is about a circle geometry theorem that claims that all angles subtended by the arc AB at the circumference of a circle are identical in size. The arc AB in figure a subtends two angles on the circumference of the circle, one at the point R and one at Q. As a result, the following is stated in this theorem: The size of the two angles ARB and AQB is the same. The theorem is not confined to only two angles, but holds true for any number of angles at the circle's circumference, as long as they are all subtended by the arc AB.

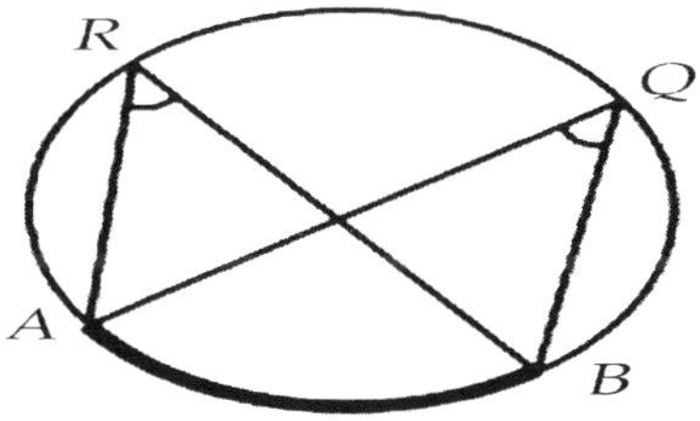

Figure 13. Angles on the same arc.

Source: Image by Wikimedia commons.

Arc

An arc is a section or sector of a curve. It could also be a segment of a line graph. The arc AB in the figure is a portion of a circle. When the arc of a circle is smaller than a semicircle, it is referred to as a minor arc; when the arc is bigger than a semicircle, it is referred to as a major arc. The arc AB of the circle in the figure is a minor arc and it is smaller than a semicircle.

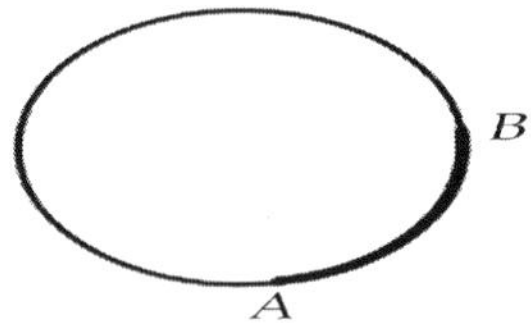

Figure 14. Arc.

Source: Image by Wikimedia commons.

Arc Length

The length of a circle's arc can be computed as shown in Figure 15.

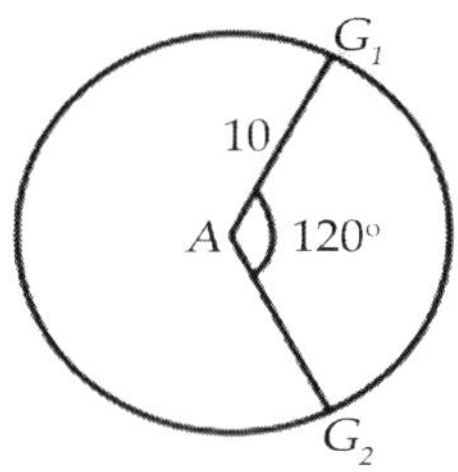

Figure 15. Arc length.

Source: Image by Wikipedia.

Area

A surface's area is a measurement of the two-dimensional space occupied by it. The unit of measurement in square units, which are denoted by the symbol units square. The area of a shape, such as a rectangle, can be calculated by counting the number of squares taken up by its surface. The area of the rectangle in figure a is calculated by counting the number of square centimeters it takes up. The abbreviation for square centimeters is cm^2. This rectangle has a surface area of 12 cm^2. Conversely, by measuring the length (4 cm) and width (3 cm) of the rectangle, the area can be computed by using the formula

Area = length × width

$A = 4 \times 3$

$= 12$

The area of this rectangle is 12 cm^2.

Finding the area of some shapes, such as circles, by counting squares is tricky since the method would require piecing together little portions of the circle to make up full squares, akin to a jigsaw. Thus, the only suitable approach is to take a formula to calculate its area.

Arithmetic

The branch of mathematics that employs numbers in operations such as addition, subtraction, multiplication, division, and square root calculation.

Arithmetic Mean

Arithmetic mean is sometimes shortened to the single word mean. It is one of three statistical averages, the other two being the median and mode. Mean is represented by the symbol $\bar{x}$. The term mean is commonly misunderstood as "average." The mean is a singular quantity that is used for expressing a set of quantities or a group of quantities. To calculate the mean of a group of quantities, add all of them together and divide the total amount by the number of quantities.

Arrow Graph

This is often referred to as an arrow diagram. Assuming that there is a "is the capital of" relationship between a set of cities and a set of countries. A pairing off may be "London is the capital of England," and an arrow would be drawn from London to England on the graph to illustrate that this relationship exists. Figure a is an arrow graph illustrating four alternative pairings for the relation "is the capital of." If the arrows' directions were changed, the outcome would be the inverse relationship: "has as its capital."

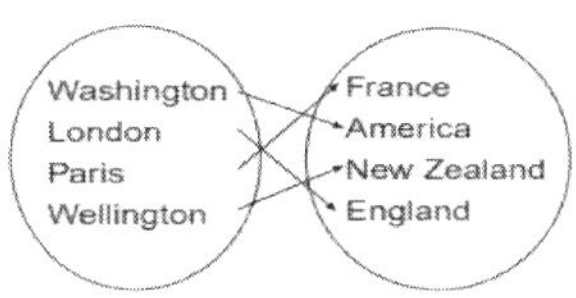

(a)

Associative Law

Associative law, in mathematics, either of two laws relating to number operations of addition and multiplication, stated symbolically: a + (b + c) =

(a + b) + c, and a(bc) = (ab)c; that is, the terms or factors may be associated in any way desired. While associativity holds for ordinary arithmetic with real or imaginary numbers, there are certain applications—such as non – associative algebras—in which it does not hold.

In Mathematics, associative law is applied to the addition and subtraction of three numbers. According to this law, if a, b and c are three numbers, then;

$a+(b+c) = (a+b)+c$

$a.(b.c) = (a.b).c$

Thus, by the above expression, we can understand that it does not matter how we group or associate the numbers in addition and multiplication. The associative law holds only for the addition and multiplication of all the real numbers but not for subtraction and division.

Also, learn:

1. Commutative Law
2. Distributive Property

In Math's, associative law is applicable to only two of the four major arithmetic operations, which are addition and multiplication. But for other arithmetic operations, subtraction and division, this law is not applied, because there could be a change in result. This is due to change in position of integers during addition and multiplication, do not change the sign of the integers.

As per associative law, if we add or multiply three numbers, then their change in position or order of numbers or arrangements of numbers, does not change the result. This law is also called associative property of addition and multiplication.

Associative Law Formula

The formula for associative law or property can be determined by its definition. As per the definition, the addition or multiplication of three numbers is independent of their grouping or association. Or we can say, the grouping or combination of three numbers while adding or multiplying them does not change the result.

Let us consider A, B and C as three numbers. Then, as per this law;

$A+(B+C) = (A+B)+C$

$A \times (B \times C) = (A \times B) \times C$

Asymmetry

Asymmetry is the absence of, or a violation of, symmetry (the property of an object being invariant to a transformation, such as reflection). Symmetry is an important property of both physical and abstract systems and it may be displayed in precise terms or in more aesthetic terms. The absence of or violation of symmetry that are either expected or desired can have important consequences for a system.

Asymptote

A straight line is an asymptote of a curve if it gets closer and closer to the curve but never quite reaches it. The straight line in figure a is an asymptote of the curve. In this example, as x increases, the curve and asymptote move closer and closer together. The gap between them is constantly narrowing.

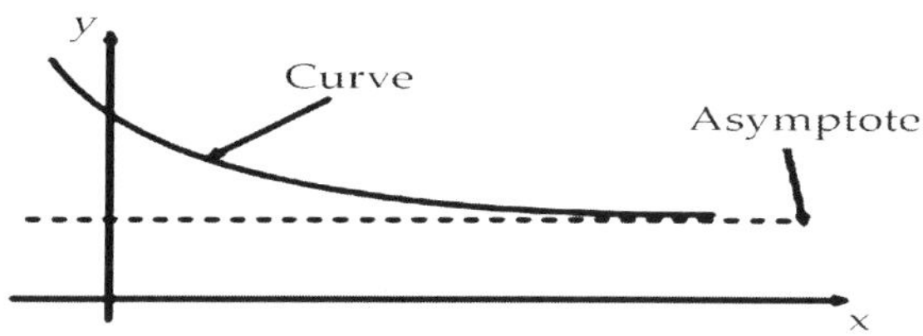

Figure 16. Asymptote.

Source: Image by Wikimedia commons.

Average

A quantity that denotes the three statistical words mode, median, and mean. The mean is sometimes referred to as the arithmetic mean, which is frequently wrongly referred to as the average.

Average Speed

The average speed of a trip is computed by dividing the entire distance travelled by the total time required to finish the journey. Any stops along the way are usually included in the overall time taken. Average speed is calculated using the formula

Average speed = total distance traveled/ total time taken

It is critical to ensure that the distance and time units are compatible. The distance must be in kilometers and the time must be in hours for an average speed measured in kilometers per hour (km/h). Other commonly used speed units are:

- Centimeters per second, cm s^{-1} or cm/s
- Meters per second, m s^{-1} or m/s
- Miles per hour, mi h^{-1} or mi/h

Axis of Symmetry

An axis of symmetry is a straight line that divides a shape into two identical halves that are mirror images of each other.

Axis of symmetry is a line that divides an object into two equal halves, thereby creating a mirror-like reflection of either side of the object. The word symmetry implies balance. Symmetry can be applied to various contexts and situations. Symmetry is a key concept in geometry which cuts the figure into two halves that are exact reflections of each other.

base n: the number of unique digits (including zero) that a positional numeral system uses to represent numbers, e.g., base 10 (decimal) uses 0, 1, 2, 3, 4, 5, 6, 7, 8 and 9 in each place value position; base 2 (binary) uses just 0 and 1; base 60 (sexagesimal, as used in ancient Mesopotamia) uses all the numbers from 0 to 59; etc.

B

Balancing An Equation

This is an equation-solving approach where the equals sign represents the equation's point of balance. To begin, the expression on the left side of the equals sign equals the expression on the right side. By performing the same operation on both sides of the equation, the balance of the equation is preserved. The use of these processes results in the solution of the equation. Addition, subtraction, multiplication, division, squaring, and finding the square root are the operations.

Bar Graph

In statistics, the bar graph, often known as a bar chart or column graph, is used for depicting data. A bar graph is made up of a series of vertical bars or rectangles of identical widths, the heights of which are proportional to the frequency of specific quantities. If the data is discrete, or divided into distinct categories, a small space is left between each bar. A histogram is used to represent continuous data. Data recorded in a frequency table are displayed considerably more efficiently in a bar graph, allowing for easier and more impactful comparisons of quantities. When creating a bar graph, it should be ensured that it has a title, that the axes are numbered, titled, and labelled, and that the heights of the columns indicate the frequencies.

Base (Geometry)

The base of a polygon is the side at the bottom of the shape. One of the polygon's sides can be turned to become the base. Assume the polygon is a triangle, for example. Based on how the triangle is drawn, each side can represent the base. Figure a depicts the same triangle in three distinct positions, demonstrating how each side in turn is the triangle's base. A triangle's altitude is its perpendicular height, or the shortest distance between the base and its highest point.

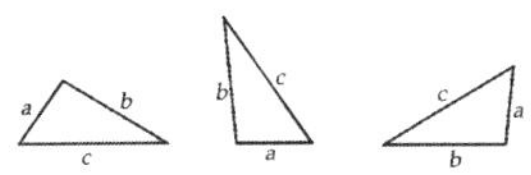

Figure 17. Base.

Source: Image by Flickr.

In the first position of the triangle in figure a, the side c is the base, in the second position side a is the base, and in the third position side b is the base. When calculating the area of a triangle it may be advantageous to carefully select one side as the base in preference to the others.

Bayesian Probability

A popular interpretation of probability which evaluates the probability of a hypothesis by specifying some prior probability, and then updating in the light of new relevant data. Bayesian probability is an interpretation of the concept of probability, in which, instead of frequency or propensity of some phenomenon, probability is interpreted as reasonable expectation representing a state of knowledge or as quantification of a personal belief.

The Bayesian interpretation of probability can be seen as an extension of propositional logic that enables reasoning with hypotheses; that is, with propositions whose truth or falsity is unknown. In the Bayesian view, a probability is assigned to a hypothesis, whereas under frequentist inference, a hypothesis is typically tested without being assigned a probability.

Bayesian probability belongs to the category of evidential probabilities; to evaluate the probability of a hypothesis, the Bayesian probability specifies a prior probability. This, in turn, is then updated to a posterior probability in the light of new, relevant data (evidence).

The Bayesian interpretation provides a standard set of procedures and formulae to perform this calculation. The term Bayesian derives from the 18th-century mathematician and theologian Thomas Bayes, who provided the first mathematical treatment of a non-trivial problem of statistical data analysis using what is now known as Bayesian inference. Mathematician Pierre-Simon Laplace pioneered and popularized what is now called Bayesian probability.

Bearings

A bearing is the direction to travel in. It is a clockwise angle measured from the true north. A bearing is a navigational term that is stated in degrees using three digits. Compass points can also be used to represent bearings.

BEDMAS

A mnemonic for remembering the sequence in which operations are carried out when performing arithmetic calculations; the order is:

1. B, Brackets
2. E, Exponents
3. D, Division
4. M, Multiplication
5. A, Addition
6. S, Subtraction

In solving complicated operations problems, BEDMAS is an incredibly useful aid. Brackets are instructed to be worked out first before the next operation takes place. It is beneficial sometimes to introduce your own brackets to simplify the order of operations further. The calculator automatically performs the operations in the correct order, if you enter these numbers and operations in the precise order in which they are stated, from left to right. Work from the inside out if more than one set of brackets are given. Prior to multiplication, the mnemonic mentions division, even though both are equal in rank. Addition and subtraction, likewise, have the same rank.

Bell Curve

The shape of the graph that indicates a normal distribution in probability and statistics.

A bell curve is a common type of distribution for a variable, also known as the normal distribution. The term "bell curve" originates from the fact that the graph used to depict a normal distribution consists of a symmetrical bell-shaped curve.

The highest point on the curve, or the top of the bell, represents the most probable event in a series of data (its mean, mode, and median in this case), while all other possible occurrences are symmetrically distributed around the mean, creating a downward-sloping curve on each side of the peak. The width of the bell curve is described by its standard deviation.

The term "bell curve" is used to describe a graphical depiction of a normal probability distribution, whose underlying standard deviations from the mean create the curved bell shape. A standard deviation is a measurement used to quantify the variability of data dispersion, in a set of given values around the mean. The mean, in turn, refers to the average of all data points in the data set or sequence and will be found at the highest point on the bell curve.

Financial analysts and investors often use a normal probability distribution when analyzing the returns of a security or of overall market sensitivity. In finance, standard deviations that depict the returns of a security are known as volatility.

For example, stocks that display a bell curve usually are blue-chip stocks and ones that have lower volatility and more predictable behavioral patterns. Investors use the normal probability distribution of a stock's past returns to make assumptions regarding expected future returns.

In addition to teachers who use a bell curve when comparing test scores, the bell curve is often also used in the world of statistics where it can be widely applied. Bell curves are also sometimes employed in performance management, placing employees who perform their job in an average fashion in the normal distribution of the graph. The high performers and the lowest performers are represented on either side with the dropping slope. It can be useful to larger companies when doing performance reviews or when making managerial decisions.

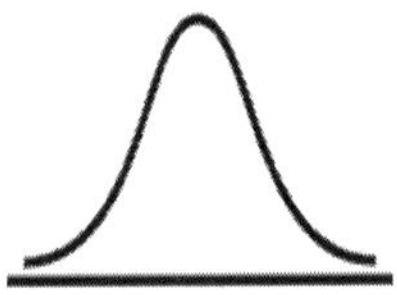

Figure 18. Bell curve.

Source: Image by Noun Project.

Bijection

A one-to-one comparison or correspondence of the members of two sets, so that there are no unmapped elements in either set, which are therefore of the same size and cardinality.

In mathematics, a bijection, bijective function, one-to-one correspondence, or invertible function, is a function between the elements of two sets, where each element of one set is paired with exactly one element of the other set, and each element of the other set is paired with exactly one element of the first set. There are no unpaired elements. In mathematical terms, a bijective function $f: X \rightarrow Y$ is a one-to-one (injective) and onto (surjective) mapping of a set X to a set Y. The term one-to-one correspondence must not be confused with one-to-one function.

Binomial

A polynomial algebraic expression or equation with just two terms, e.g., $2x^3 - 3y = 7$; $x^2 + 4x$; etc.

The binomial is a type of distribution that has two possible outcomes (the prefix "bi" means two, or twice). For example, a coin toss has only two possible outcomes: heads or tails and taking a test could have two possible outcomes: pass or fail. A Binomial Distribution shows either Success or Failure.

Binomial Coefficients

The coefficients of the polynomial expansion of a binomial power of the form $(x + y)^n$, which can be arranged geometrically according to the Binomial theorem as a symmetrical triangle of numbers known as Pascal's Triangle, e.g., $(x + y)^4 = x^4 + 4x^3y + 6x^2y^2 + 4xy^3 + y^4$ the coefficients are 1, 4, 6, 4, 1

Boolean Algebra or Logic

A type of algebra which can be applied to the solution of logical problems and mathematical functions, in which the variables are logical rather than numerical, and in which the only operators are AND, OR and NOT

Box Plot

The method to summarize a set of data that is measured using an interval scale is called a box and whisker plot. These are maximum used for data analysis. We use these types of graphs or graphical representation to know:

- Distribution shape.
- Central value of it.
- Variability of it.

A box plot is a chart that shows data from a five-number summary including one of the measures of central tendency. It does not show the distribution in particular as much as a stem and leaf plot or histogram does. But it is primarily used to indicate a distribution is skewed or not and if there are potential unusual observations (also called outliers) present in the data set.

Boxplots are also very beneficial when large numbers of data sets are involved or compared. In simple words, we can define the box plot in terms of descriptive statistics related concepts. That means box or whiskers plot is a method used for depicting groups of numerical data through their quartiles graphically. These may also have some lines extending from the boxes or whiskers which indicates the variability outside the lower and upper quartiles, hence the terms box-and-whisker plot and box-and-whisker diagram. Outliers can be indicated as individual points.

It helps to find out how much the data values vary or spread out with the help of graphs. As we need more information than just knowing the measures of central tendency, this is where the box plot helps. This also takes less space. It is also a type of pictorial representation of data.

Since, the center, spread and overall range are immediately apparent, using these boxplots the distributions can be compared easily.

Parts of Box Plots

Minimum: The minimum value in the given dataset

First Quartile (Q1): The first quartile is the median of the lower half of the data set.

Median: The median is the middle value of the dataset, which divides the given dataset into two equal parts. The median is considered as the second quartile.

Third Quartile (Q3): The third quartile is the median of the upper half of the data.

Maximum: The maximum value in the given dataset.

Apart from these five terms, the other terms used in the box plot are:

Interquartile Range (IQR): The difference between the third quartile and first quartile is known as the interquartile range. (i.e.) IQR = Q3-Q1

Outlier: The data that falls on the far left or right side of the ordered data is tested to be the outliers. Generally, the outliers fall more than the specified distance from the first and third quartile.

(i.e.) Outliers are greater than Q3+(1.5. IQR) or less than Q1–(1.5. IQR).

Bias

A word for describing an unfair population sample used in statistics. A biased sample is one that does not represent the selected population accurately. The prediction of the population from which it was collected cannot be made by using a biased sample. It is vital to determine why a sample is biased and to be able to choose another sample that is not biased, so that sample mistakes do not reoccur.

There is a famous incident about the survey made by an American magazine shortly before the 1936 presidential election. A massive sample of 2 million people was drawn from the magazine's readers, who were then contacted via phone.

The survey results expected Franklin D. Roosevelt to be enormously defeated. In reality, the opposite occurred, and Roosevelt was re-elected with a majority. The sample was partial because the sample was only available for the reader and only for customers of the telephone. For a number of reasons, a sample can be biased.

Billion

This figure is commonly used, particularly in the United States, as 1000 million, that is 1,000,000,000, or 109. Other countries, such as the United Kingdom and Germany, consider it to be 1012, or one million.

Bimodal

A collection of data is said to be bimodial when it has two distinct modes, and the frequency curve of the distribution has two "humps."

Binary Digit

A binary digit can only be a 0 or a 1. Frequently shortened as "bit."

Binary Numbers

Numbers written in base 2, as opposed to numbers in common use written in base 10, which are known as denary numbers. Assume there is a set of weights in grams (abbreviated g) that is used for weighing certain quantities. The weights are as follows: 1, 2, 4, 8, and 16 g.

The table displays the weights required to measure various quantities. Binary numbers include the numbers 1, 10, 11, 1000, 1100, 1111, 10001, and 10011, which represent the number of weights required to weigh specific quantities. As shown in the following example, binary numbers can be summed together.

Number of Each Weight Needed					Quantity to be Weighed
16 g	8 g	4 g	2 g	1 g	
				1	1 g
			1	0	2 g
			1	1	3 g
	1	0	0	0	8 g
	1	1	0	0	12 g
	1	1	1	1	15 g
1	0	0	0	1	17 g
1	0	0	1	1	19 g

Figure 19. Binary numbers.

Source: Image by Wikimedia commons.

Bisect

Bisecting a geometrical figure refers to dividing it into two congruent or equal sections. Geometrical constructions for perpendicular bisection of a line segment using a ruler and compasses are detailed in the entry Perpendicular Bisector, and bisection of an angle is explained in the entry Angle Bisector. The resulting construction line is a mirror line for both instances.

Box and Whisker Graph

This graph is also known as a box plot. Commonly used in statistics to show the distribution of data around the median. The following are the components of box and whisker graphs:

1. Greatest value
2. Upper quartile, UQ
3. Median, M
4. Lower quartile, LQ
5. Least value
6. Range
7. Interquartile range

Braces

Braces are a form of bracket; denoted by the symbol { }. They are used to signify a set, and the set's elements are given inside the braces. For example, the winter sports available at Luke's Community Center include {baseball, basketball, hockey, and soccer}.

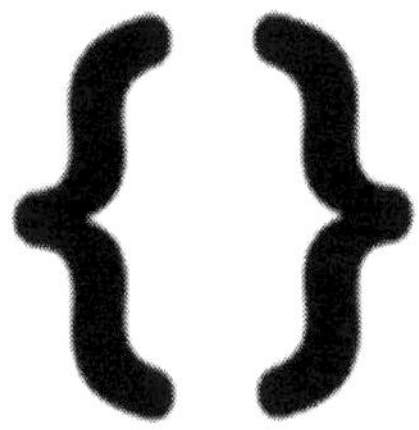

Figure 20. Braces.

Source: Image by Noun Project.

Brackets

Brackets are used to encapsulate something and is a common term for the following:

1. Braces { }, used to enclose a set-in set theory. Refer to the entry Braces for an example.
2. Curved brackets ( ), also known as parentheses in algebra, are used to surround terms that indicate a single quantity or expression.

Breadth

Width is another, more frequent word for breadth. A geometrical shape's breadth is given as the distance measured across the shape at its widest point in a direction perpendicular to the shape's length. A shape's length is its longest measurement. The units for estimating breadth are the same as those for measuring length: millimeters, centimeters, meters, and kilometers. In the formula for calculating the area of a rectangle, the word breadth may be substituted for the word width.

C

Calculate

Calculation or to calculate is the process of using numbers to obtain the answer to a question. When answering an algebraic expression question, the word simplify is used as the command instead of calculate.

Calculator

A scientific calculator with direct algebraic logic, abbreviated DAL, is commonly used to help study mathematics. DAL refers to entering information into the calculator in the same order that it would normally be written down in. Considering that different calculators have slightly varied instructions; it was decided not to include the calculator methods in this book. Instead, the calculator manual can be referred to.

Calculus (Infinitesimal Calculus)

A branch of mathematics involving derivatives and integrals, used to study motion and changing values. A term which formerly included various branches of mathematical analysis connected with the concept of an infinitely-small function. Even though the method of "infinitely smalls" had been successfully employed in various forms by the scientists of Ancient Greece and of Europe in the Middle Ages to solve problems in geometry and in natural science, exact definitions of the fundamental concepts of the theory of infinitely-small functions were laid only in the 19th century. In order to grasp the importance of this method, it must be pointed out that it was not the infinitesimal calculus itself which was of practical importance, but only the cases in which its use resulted in finite quantities.

Calculus of Variations

An extension of calculus used to search for a function which minimizes a certain functional (a functional is a function of a function).

Calculus of variations, branch of mathematics concerned with the problem of finding a function for which the value of a certain integral is either the largest or the smallest possible. Many problems of this kind are easy to state, but their solutions commonly involve difficult procedures of the differential calculus and differential equations.

The isoperimetric problem—that of finding, among all plane figures of a given perimeter, the one enclosing the greatest area—was known to Greek mathematicians of the 2nd century BCE. The term isoperimetric problem has been extended in the modern era to mean any problem in the calculus of variations in which a function is to be made a maximum or a minimum, subject to an auxiliary condition called the isoperimetric condition, although it may have nothing to do with perimeters. For example, the problem of finding a solid of given volume that has the least surface area is an isoperimetric problem, the given volume being the auxiliary, or isoperimetric, condition. An example of an isoperimetric problem from the field of aerodynamics is that of finding the shape of a solid having a given volume that will encounter minimum resistance as it travels through the atmosphere at a constant velocity.

Modern interest in the calculus of variations began in 1696 when Johann Bernoulli of Switzerland proposed a brachistochrone ("least-time") problem as a challenge to his peers.

Cardinal Numbers

Numbers used to measure the cardinality or size (but not the order) of sets – the cardinality of a finite set is just a natural number indicating the number of elements in the set; the sizes of infinite sets are described by transfinite cardinal numbers, (aleph-null), (aleph-one), etc.

A Cardinal Number describes or represents how many of something are present. Example 2 apples, 5 flowers, etc. It quantifies an object. It does not have values as fractions or decimals. Cardinal numbers are counting numbers, they help to count the number of items. Let's have a look at cardinal numbers examples. Ana wants to count the number of people standing in a queue at a billing counter. Can you help her? Ana started to count using Natural numbers.

Ana counted 1, 2, 3, 4, and 5. There are 5 people standing in a queue at the billing counter. Counting numbers are cardinal numbers! Now, Let's consider another example, Noah kept eight apples in a basket. The number eight denotes how many apples are there in the basket, irrespective of their order.

Examples of cardinal numbers are 1, 2, 3, 4, 5, 6, 7, 8, 9, 10, 11, 12, 13, 14, 15, 16, 17, 18, 19, 20, 21, and so on. The smallest Cardinal number is 1 as 0 is not used for counting, so it is not a cardinal number.

Figure 21. Cardinal number.

Source: Image by Pixabay.

Canceling

Canceling down of fractions refers to rewriting them in their simplest form as equivalent fractions.

Capacity

The volume of liquid that a container can hold is referred to as its capacity. The capacity units are given below:

1. Milliliters (ml). One milliliter is equal to one cubic centimeter in volume (cm^3, or cc). It has about the same capacity as a teaspoon.
2. Liter (l). 1000 milliliters equal one liter. It has the capacity of a large jug, or approximately 9/2 cups.
3. Kiloliter (kl). A kiloliter equals 1000 liters. It is the capacity of a cube with a volume of one cubic meter ($1\ m^3$).

Capital

An amount of money that a person requires in order to start a business or money that a person has collected that can be invested profitably.

Cartesian Coordinates

This is a method for determining the position of a point in a plane based on its distances from two perpendicular number lines known as axes. An excellent method to demonstrate how Cartesian coordinates function is using an example.

Treasure Island is depicted on a map with an East axis and a North axis. These axes are labelled with numbers at equal intervals. The position of any point on the map can be determined by specifying its number on the East axis as well as its number on the North axis. The East number for Quick Sands is 2 and the North number is 4.

This Quick Sands reference is abbreviated as (2, 4), with the East number written before the North number. With respect to these axes, the coordinates of the point Quick Sands are (2, 4). The point at which the two axes intersect is referred to as the origin, and instead of two zeros, one zero is used for denoting the point of intersection.

Rene Descartes (1596–1650) devised the grid system, which consists of two axes drawn on a grid of squares and takes his name: "Cartesian." He labelled the East axis as the x-axis and the North axis as the y-axis, and he incorporated negative integers on both the axes. The two numbers in a bracket that indicate a point, such as Quick Sand (2, 4), are known as ordered pairs, or a point's coordinates. The first value in the bracket, 2, is known as the x coordinate, while the second number, 4, is known as the y coordinate. Plotting a point is the act of marking the location of a point on a set of axes.

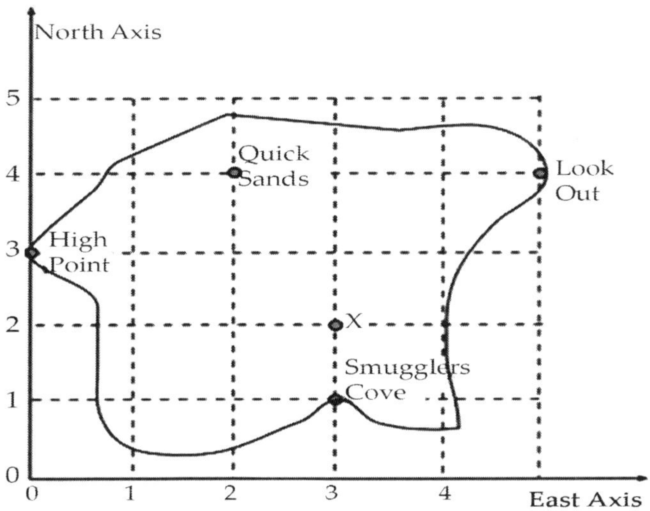

Figure 22. Cartesian coordinates.

Source: Image by Wikimedia commons.

Celsius

A unit used for measuring temperature. It was originally as centigrade meaning "100 steps."

Census

The formal counting of all the people in a population while also gathering additional information about them including religion, age, gender, income, possessions, and so on. Every few years, countries undertake a population census. A survey is a type of research that is conducted on a sample of a population. A survey can take the form of distributing a questionnaire, measuring a quantity, or counting items.

A survey is often conducted on a subset of a population, and the data gained is used to make predictions and draw inferences about the population. A municipal council, for example, can conduct a survey of sample of its inhabitants on where to a new library can be constructed, using the results to determine the opinion of all city residents.

Centi

A prefix that implies one-hundredth of something. 1/100 is one hundredth written as a fraction. A centimeter, for example, is one-hundredth of a meter and is shortened as cm, where c represents centi and m represents meter. Since metric units are primarily based on dividing (or multiplying) a number by 1000, the prefix centi is rarely used except in centimeter and cent, which is 1/100 of a dollar.

Central Tendency

The three averages mean, mode, and median have a propensity to be towards the center of a normal distribution that has been sorted in order of size. They are referred to as measures of central tendency for the distribution. The more data there is, the more possible it is that the three averages are closer to each other in value and closer to the center.

The three averages are not expected to be around the center of the ranked data for a small amount of data that is not regularly distributed. Studying central tendency provides an excellent opportunity to explore the three averages, mean, mode, and median, as representative values. They can be utilized in depicting a quantity of data so that the data can be compared with another.

Cgs System of Units

This measuring system employed centimeters (cm) to measure length, grams (g) to measure mass and seconds (s) to measure time on the basis of the metric system. In 1960 the Systeme International d'Unites or the SI units had substituted these units.

The SI units are used as basic elements of the length, mass, and time as meters (m), kilogram me (kg) and seconds(s). Multiple and fractions of 1000 of the basic units are some other units in use in this system. Exceptions to the 1000 multiple (and fractional) rule are 10 mm = 1 cent, 100 mm = 1 meter, 10000 m^2 = 1 hectare, and 100 hectares = 1 km^2.

Chance

Another term for probability.

Changing the Subject of A Formula

If the radius of a circle is given, then the formula for calculating the area of a circle is given as $A = \pi R^2$. A, the term on the left side of the equals sign, is the subject of this formula. Assume the area of a circle is known and the radius has to be. In other words, a formula in which R is the subject is required.

Changing the subject of a formula refers to the process of rearranging a formula so that another variable becomes the subject. The rules for changing a formula's subject and the rules for solving an equation are the same. The first step in modifying the subject of a formula is rewriting the formula with the new-subject on the left side of the equality sign.

Chord

A chord is a line segment that connects two points lying on a curved line. The circle is the curve under study. Figure a shows that AB is a chord of the circle. The center of the circle always lies on the perpendicular bisector or mediator of a circle's chord. The diameter is called the longest feasible chord, passing through the center O of the circle. The secant of a circle is obtained when one or both ends of a chord are stretched. Figure c shows two examples.

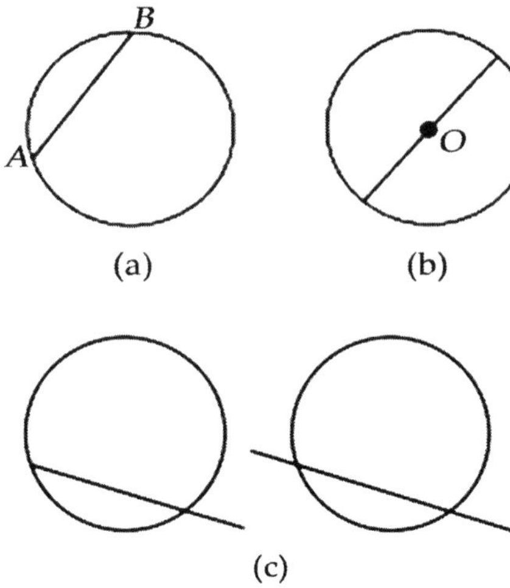

Figure 23. Chord.

Source: Image by Wikipedia.

Circle

A collection of points in a plane, all of which are at an equal distance from a fixed-point O. The fixed point is known as the circle's center.

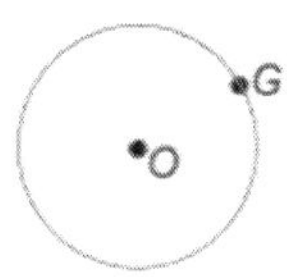

Figure 24. Circle.

Source: Image by Wikimedia commons.

Circumcenter

This is the center of the circle passing through the three vertices of a triangle or any polygon's vertices. This circle is known as the triangle ABC's circumcircle. If the angle is obtuse, the circumcenter can lie outside the triangle. A circumcircle can be constructed through the vertices of any triangle, although this is not always the case for other polygons.

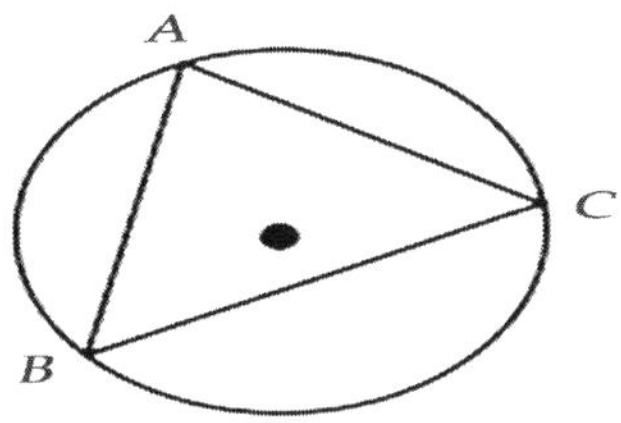

Figure 25: Circumcenter.

Source: Image by Wikimedia commons.

Circumference

The circumference of a circle refers to the circle's curved border line or the length of the circle's boundary. It is a term that is frequently used to describe various closed geometrical shapes. A circle's circumference (C) can be computed using the formula $C = \pi D$, where D is the circle's diameter. The entry Circle has an example calculating the circumference of a circle.

Circumscribe

This involves drawing a circle, or any closed curve, around the exterior of a polygon such that the circle goes through all of the polygon's vertices. Figure a depicts a square peg fitting neatly into a round hole. The square peg is surrounded or circumscribed by a round hole. The square's sides are chords of the circle. Inscribe refers to drawing a circle within a polygon so that all of the polygon's sides slightly touch the circle. Figure b depicts a view of a round cake in a square box from the top. The square has a circle inscribed in it. The square's sides are tangents to the circle.

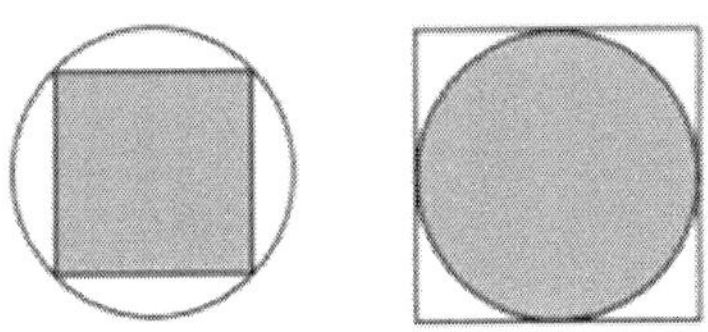

Figure 26. Circumscribe.

Source: Image by Flickr.

Coefficient

In an algebraic term, the coefficient of a variable is the integer that multiplies the variable. In the term 3x, the coefficient of the variable x is 3, as 3 multiplies the variable x. Here are some more examples:

- The coefficient of x^2y in the term $-5x^2y$ is -5.
- The coefficient of xy in the term 3x + 5xy is 5, and the coefficient of x is 3.
- The coefficient of xy in the term 2 + xy is 1, because 1xy is written as xy.

Collecting Data

Collecting data refers to the process of gathering and recording of information, which is often numerical, so that it can be evaluated and conclusions can be drawn from it, and it can also be used to help make decisions or predictions. Data is collected in response to an issue, and the method used to collect the data varies depending on what the problem is.

Collinear

If all of the points in a set are on the same straight line, the set is said to be collinear. A straight line can always be drawn from two points; however, this is not always true for three or more points.

Column

In a table or vector, a column refers to a vertical or erect line of numbers or phrases. A row is perpendicular to a column.

Column Graph

This is similar to a bar graph, however here the bars are positioned vertically as columns, while bar graph has bars can be drawn either vertically or horizontally.

Combinations

Combinations are sometimes known as selections, which is a more apt term to describe the process. A selection is the numerous ways a set of r elements can be selected from a set of n elements when the order of the elements is ignored.

Combinatorics

The study of different combinations and groupings of numbers, often used in probability and statistics, as well as in scheduling problems and Sudoku puzzles.

Combinatorics is a branch of mathematics which is about *counting* – and we will discover many exciting examples of "things" you can count.

First combinatorial problems have been studied by ancient Indian, Arabian and Greek mathematicians. Interest in the subject increased during the 19th and 20th century, together with the development of graph theory and problems like the four-color theorem. Some of the leading mathematicians include Blaise Pascal (1623 – 1662), Jacob Bernoulli (1654 – 1705) and Leonhard Euler (1707 – 1783).

Combinatorics has many applications in other areas of mathematics, including graph theory, coding and cryptography, and probability.

Complex Dynamics

The study of mathematical models and dynamical systems defined by iteration of functions on complex number spaces. The dynamics alluded to by the title of the course refers to dynamical systems that arise from iterating a holomorphic self-map of a complex manifold. In this course, the manifolds underlying these dynamical systems will be of complex dimension 1.

The foundations of complex dynamics are best introduced in the setting of compact spaces. Iterative dynamical systems on compact Riemann surfaces other than the Riemann sphere – viewed here as the one-point compactification of the complex plane – are relatively simple. We shall study what this means. Thereafter, the focus will shift to rational functions: these are the holomorphic self-maps of the Riemann sphere. Along the way, some of the local theory of fixed points will be presented.

In the case of rational maps, some ergodic-theoretic properties of the orbits under iteration will be studied. The development of the latter will be self-contained. The properties/ theory covered will depend on the time available and on the audience's interest.

Complex Number

A number expressed as an ordered pair comprising a real number and an imaginary number, written in the form $a + bi$, where a and b are real numbers, and i is the imaginary unit (equal to the square root of -1)

Composite Number

A number with at least one other factor besides itself and one, i.e., not a prime number.

Numbers can be classified on the basis of the number of factors they have. If a number has just two factors – 1 and the number itself, then it is a prime number. However, most numbers have more than two factors, and they are called composite numbers.

Composite numbers can be defined as natural numbers that have more than two factors. In other words, a number which is divisible by a number other than 1 and the number itself, is called a composite number.

A composite number is a positive integer that can be formed by multiplying two smaller positive integers. Note the properties of a composite number listed below:

1. All composite numbers are evenly divisible by smaller numbers that can be prime or composite.
2. Every composite number is made up of two or more prime numbers.

The two main types of composite numbers in math's are odd composite numbers and even composite numbers. Let us have a look at the two of them individually:

Odd Composite Numbers

All the odd numbers which are not prime are odd composite numbers. For example, 9, 15, 21, 25, 27 are odd composite numbers. Consider the numbers 1, 2, 3, 4, 9, 10, 11, 12 and 15. Here 9 and 15 are the odd composites because these two numbers have odd divisors and satisfy the composite condition.

Even Composite Numbers

All the even numbers which are not prime are even composite numbers. For example, 4, 6, 8, 10, 12, 14, 16, are even composite numbers. Consider the numbers 1, 2, 3, 4, 9, 10, 11, 12 and 15 again. Here 4, 10, and 12 are the even composites because they have even divisors and satisfy the composite condition.

Smallest Composite Number

A composite number is defined as a number that has divisors other than 1 and the number itself. Start counting: 1, 2, 3, 4, 5, 6, ... so on. 1 is not a composite number because its sole divisor is 1. 2 is not a composite number because it has only two divisors, i.e., 1 and the number 2 itself. 3 is not a composite number because it has only two divisors, i.e., 1 and the number 3 itself. Let's look at number 4. Its divisors are 1, 2, and 4. Number 4 satisfies the criteria of a composite number. So, 4 is the smallest composite number.

Congruence

Two geometrical figures are congruent to one another if they have the same size and shape, and so one can be transformed into the other by a combination of translation, rotation and reflection.

Congruence, in mathematics, a term employed in several senses, each connoting harmonious relation, agreement, or correspondence.

Two geometric figures are said to be congruent, or to be in the relation of congruence, if it is possible to superpose one of them on the other so that they coincide throughout. Thus, two triangles are congruent if two sides and their included angle in the one is equal to two sides and their included angle in the other. This idea of congruence seems to be founded on that of a "rigid body," which may be moved from place to place without change in the internal relations of its parts.

The position of a straight line (of infinite extent) in space may be specified by assigning four suitably chosen coordinates. A congruence of lines in space is the set of lines obtained when the four coordinates of each line satisfy two given conditions. For example, all the lines cutting each of two given curves form a congruence.

The coordinates of a line in a congruence may be expressed as functions of two independent parameters; from this it follows that the theory of congruences is analogous to that of surfaces in space of three dimensions. An important problem for a given congruence is that of determining the simplest surface into which it may be transformed.

Two integers a and b are said to be congruent modulo m if their difference $a–b$ is divisible by the integer m. It is then said that a is congruent to b modulo m, and this statement is written in the symbolic form $a \equiv b$ (mod m). Such a relation is called a congruence. Congruences, particularly those involving a variable x, such as $xp \equiv x$ (mod p), p being a prime number, have many properties analogous to those of algebraic equations. They are of great importance in the theory of numbers.

Conic Section

The section or curve formed by the intersection of a plane and a cone (or conical surface), depending on the angle of the plane it could be an ellipse, a hyperbola or a parabola

Continued Fraction

A fraction whose denominator contains a fraction, whose denominator in turn contains a fraction, etc.

Coordinate

The ordered pair that gives the location or position of a point on a coordinate plane, determined by the point's distance from the x and y axes, e.g., (2, 3.7) or (–5, 4).

Coordinates are a set of two values that locate a specific point on a coordinate plane grid, better known as a coordinate plane. A point in a coordinate plane is named by its ordered pair of the form (x, y), which is written in parentheses, corresponding to the X-coordinate and the Y-coordinate. These coordinates can be positive, zero, or negative, depending on the location of the point in the respective quadrants.

Coordinate Plane

A plane with two scaled perpendicular lines that intersect at the origin, usually designated x (horizontal axis) and y (vertical axis).

A coordinate plane is a two-dimensional plane formed by the intersection of a vertical line called the y-axis and a horizontal line called the x-axis. These are perpendicular lines that intersect each other at zero, and this point is called the origin.

A coordinate plane is a two-dimensional surface formed by two number lines. It is formed when a horizontal line called the X-axis and a vertical line called a Y-axis intersect at a point called an origin. The numbers on a coordinate grid are used to locate points. A coordinate plane can be used for graph points, lines, and much more. It acts as a map and yields precise directions from one point to another. The coordinate plane definition is as follows: A coordinate plane, also known as a rectangular coordinate plane grid, is a two-dimensional plane formed by the intersection of a vertical line called Y-axis and a horizontal line called X-axis.

Correlation

A measure of relationship between two variables or sets of data, a positive correlation coefficient indicating that one variable tends to increase or decrease as the other does, and a negative correlation coefficient indicating that one variable tends to increase as the other decreases and vice versa.

Correlation refers to the degree of correspondence or relationship between two variables. Correlated variables tend to change together. If one variable gets larger, the other one systematically becomes either larger or smaller.

The correlation coefficient tells us that a relationship exists. The + or – sign indicates the direction of the relationship while the number indicates the magnitude of the relationship. This relationship should not be interpreted as a causal relationship. Variable X is related to variable Y, and may indeed be a good predictor of variable Y, but variable X does not cause variable Y although this is sometimes assumed. For example, there may be a positive correlation between head size and IQ or shoe size and IQ. Yet no one would say that the size of one's head or shoe size causes variations in intelligence. However, when two more likely variables show a positive or negative correlation, many interpret the change in the second variable to have been caused by the first.

Center of Enlargement

The position of the enlarged shape is described by the center of enlargement (O in the following diagrams).

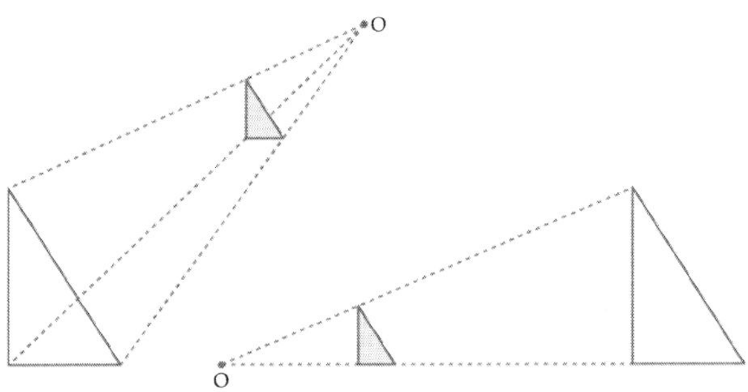

Figure 27. Center of enlargement.

Source: Image by Wikimedia commons.

Center of Rotation

The center of rotation is a point about which a plane figure rotates. This point does not move during the rotation.

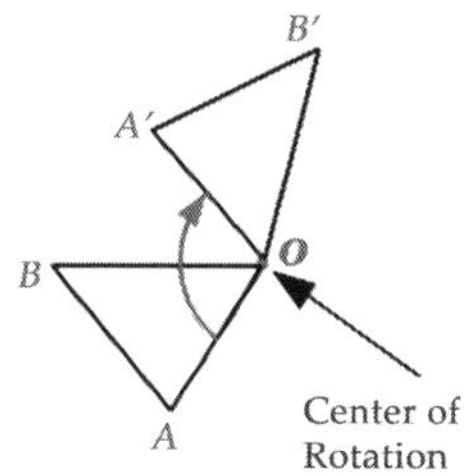

Figure 28. Center of rotation.

Source: Image by Wikimedia commons.

Centigrade

1. Centigrade is a measure of temperature.
2. Centigrade is represented by the symbol '°C.'
3. Another name for Centigrade is 'Celsius.'

The freezing point of water is 0°C and the boiling point of water is 100°C under a pressure of one standard atmosphere.

Centimeter

A centimeter is a metric unit for the measurement of length of objects and small distances. It is written using the symbol cm. It can also be defined as the unit of length in the International System of Units (SI), the current form of the metric system. It is equivalent to 1/100 meters.

A centimeter is based on the concept meter, SI unit, and as the prefix "centi" denotes that is equal to one-hundredth of a meter. Based on the decimal system, the range of the metric is the factors from 10^{-18} to 10^{18}, with the base having no prefix and 1 as the factor. Here, the base is the meter. Also, 1 cm is equivalent to 0.39370 inches.

The three conventional tools of measurement are namely ruler, meter stick and tape.

1. **Ruler:** A ruler is the most commonly used geometric tool for measurement in mathematics. It is used to measure the length

or height of small things such as notebooks, pens, pencils and so on. Most of the rulers contain two measurement metrics namely centimeters and inches. Also, these rulers have lengths of 15 cm and 30 cm in most of the cases.

2. **Meter Stick:** A measurement tool that measures one meter, i.e., 100 centimeters and is used to measure objects in meters and centimeters is called a meter stick. For example, the length of a small rope, height of a table or the width of a room, etc., can be measured with the help of a meter stick.
3. **Tape:** A tape (measuring tape) is a flexible ruler that measures the length of objects which are curative in nature such as ribbons, cloths, rings and so on.

Centimeter Scale

The scale for centimeters measurement can be easily understood with the help of a ruler. The figure below shows the measurement of length in centimeters.

This measurement is used to measure small distances between objects or dimensions of small objects in real-life situations.

Centimeter Chart

Different charts can be constructed for showing the conversion of units from centimeters to other units of measurement such as meters, inches, feet and so on. Using these charts, one easily remembers the conversion of units from centimeters to other units.

How many centimeters in one meter

As we know, a centimeter is equivalent to one-hundredth of a meter. Thus, there will be 100 centimeters in one meter.

100 cm = 1 m

Or

1 cm = 1/100 m

Circle Graph

A circle graph is also known as a pie chart. The graph is in the shape of a circle with different wedges that each represent a percentage of a total. These wedges often look like pieces of pie, which is why the circle graph is sometimes referred to as a pie chart. Each angle of the circle graph is proportional to the quantity it represents. The circle graph was first used in 1801 by William Playfair; however, it did not gain popularity until the late 1800's when used by Florence Nightingale.

There are different types of circle graphs. They might look different, but they all represent data in much the same manner.

1. **Circle Graph** This is the basic circle graph. This example shows what percentage of people belong to the world's major religions.
2. **Exploded Circle Graph** In this type of chart, the sections are 'exploded out.' This is done to show details of smaller sections or to highlight one section.
3. **3D Circle Graph** Often used for aesthetic reasons, this 3-dimensional circle graph often makes the data more difficult to interpret, but as you can see, it does indeed look pretty.
4. **Donut Chart** Functionally the same as a circle graph, but with a hole in the center, this type of graph is able to support multiple statistics as one.
5. **Exploded Donut Chart** This chart is the same as a donut chart, but the wedges have 'exploded out,' much like the Exploded Circle Graph.

Circle Theorems

Circle theorem includes the concept of tangents, sectors, angles, the chord of a circle and proofs. A circle is the locus of all points in a plane which are equidistant from a fixed point. The fixed point is called the center of the circle, and the constant distance between any point on the circle and its center is called the radius. The perimeter of a circle is known as the circumference and the area occupied by a circle in a plane is its area. The tangent is perpendicular to the radius, at any point of a circle, through the point of contact. Now for the theorems:

1. The angle at the center is twice the angle at the circumference
2. The angle in a semicircle is a right angle
3. Angles in the same segment are equal
4. Opposite angles in a cyclic quadrilateral sum to 180°
5. The angle between the chord and the tangent is equal to the angle in the alternate segment

Circumcircle

When a polygon is enclosed in a circle that passes through all of its vertices, we call that circle the circumcircle of the polygon. For example, since the circular entire risk area passes through each of the vertices of the square high-risk area, we would say that the circle is the circumcircle of the square.

Circumcircles have various characteristics and properties that make them very interesting, and provide formulas for analyzing their different aspects.

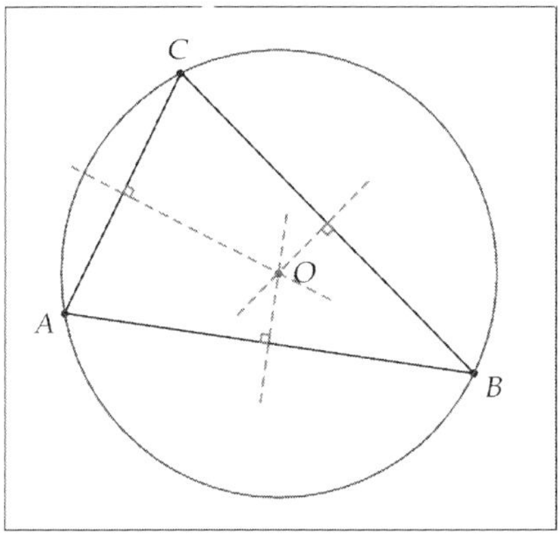

Figure 29. Circumcircle.

Source: Image by Wikimedia Commons.

Properties and Formulas

Like any circle, a circumcircle has a center point and a radius. We call the center point the **circumcenter** of the polygon that the circumcircle belongs to. The radius is a line segment from the circumcenter to any point on the circumcircle, and is called the **circumradius** of the polygon that the circumcircle belongs to.

The area and perimeter of a circumcircle are the same as they would be for any other circle. If a circle has radius r, then the formulas for the area and perimeter of that circle, are as follows:

1. Area of a circle = πr^2
2. Perimeter of a circle = $2\pi r$

We see that both of these formulas depend on the radius of a circle, so in the case of a circumcircle, the area and perimeter of a circumcircle both depend on the circumradius of the polygon that the circumcircle belongs to. Therefore, to find the area or the perimeter of a circumcircle of a polygon, we follow these steps:

1. Find the length of the circumradius of the polygon.
2. Plug the value you found in step 1 in for r in the appropriate formula.

Obviously, the circumradius of a polygon is going to depend on the type of polygon, so it will be different in each case.

Class Interval

One of the ranges into which data in a frequency distribution table (or histogram) are binned. The ends of a class interval are called class limits, and the middle of an interval is called a class mark.

Class interval refers to the numerical width of any class in a particular distribution.

Mathematically it is defined as the difference between the upper-class limit and the lower-class limit

Class interval = Upper class limit – Lower class limit

In statistics, the data is arranged into different classes and the width of such class is called class interval. Class intervals are generally equal in width but this might not be the case always.

Also, they are generally mutually exclusive. Class Intervals are very useful in drawing histograms.

Cointerior Angles

- A set of angle pairs formed when parallel lines are intersected by a transversal.
- Co-interior angles lie inside the parallel lines on the same side of the transversal.
- Also called interior angles.

Co-interior angles lie between two lines and on the same side of a transversal.

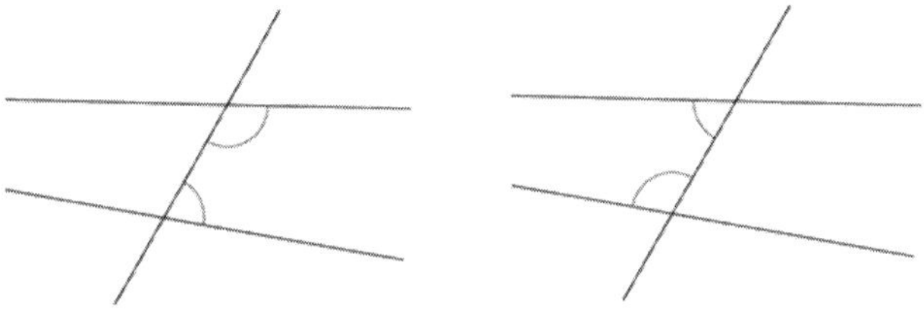

Figure 30. Cointerior angles.

Source: Image by Flickr.

In each diagram the two marked angles are called co-interior angles.

If the two lines are parallel, then co-interior angles add to give 180° and so are supplementary. In the diagram below the angles ∠CGF and ∠AFG are supplementary.

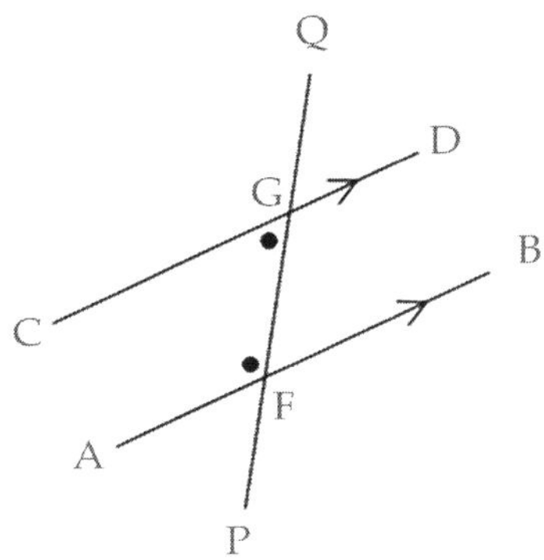

Figure 31. The angles ∠CGF and ∠AFG are supplementary to each other.

Source: Image by Noun Project.

Conversely, if a pair of angles are supplementary, then the lines are parallel. Line segment CD is parallel to line segment AB, because ∠CGF + ∠AFG = 180°.

Compass Points

The 4 main points are North, South, East and West (going clockwise they are NESW).

Halfway between each of these is North-East, South-East, South-West and North-West.

And in between all of those are:

1. NNE (north-north-east),
2. ENE (east-north-east),
3. ESE (east-south-east),
4. SSE (south-south-east),
5. SSW (south-south-west),
6. WSW (west-south-west),
7. WNW (west-north-west),
8. NNW (north-north-west).

Concyclic Points

A set of points is said to be **concyclic** if a circle passes through all of them. We have already seen that three non-collinear points are always concyclic, because there will always exist a (unique) circle passing through them. The question now is: suppose that there are four points A,B,C and D (any three of which are non-collinear). Will they always be concyclic? A little thinking will show that the answer is: No. If you draw the circle passing through, for example, A,B and C, it is not at all necessary that this same circle will pass through D as well. Therefore, four points will in general *not* be concyclic. However, there can be special situations when four points are concyclic. The next theorem discusses one such special situation.

Constant of Proportionality

Constant of proportionality is the constant value of the ratio between two proportional quantities. Two varying quantities are said to be in a relation of proportionality when, either their ratio or their product yields a constant. The value of the constant of proportionality depends on the type of proportion between the two given quantities: Direct Variation and Inverse Variation.

1. **Direct Variation:** The equation for direct proportionality is $y = kx$, which shows as x increases, y also increases at the same rate. Example: The cost per item(y) is directly proportional to the number of items(x) purchased, expressed as $y \propto x$
2. **Inverse Variation:** The equation for the indirect proportionality is $y = k/x$, which shows that as y increases, x decreases and vice-versa. Example: The speed of a moving vehicle (y) inversely varies as the time taken (x) to cover a certain distance, expressed as $y \propto 1/x$

In both the cases, k is constant. The value of this constant is called the coefficient of proportionality. The constant of proportionality is also known as unit rate.

We use constant of proportionality in mathematics to calculate the **rate of change** and at the same time determine if it is direct variation or inverse variation that we are dealing with. Let us assume that the cost of 2 apples = $20. We determine that the cost of 1 apple = $10. We have found the Constant of Proportionality for the cost of an apple is 2.

If we want to draw a picture of the Taj Mahal by sitting in front of it on a piece of paper by looking at the real image in front of us, we should maintain a proportional relationship between the measures of length, height, and width of the building. We need to identify the constant of proportionality to get the desired outcome. Based on this, we can draw the monument with proportional measurements. For instance, if the height of the dome is 2 meters then in our drawing we can represent the same dome with height 2 inches. Similarly, we can draw other parts. In such scenarios, we use constant of proportionality.

Working with proportional relationships allows one to solve many real-life problems such as:

1. Adjusting a recipe's ratio of ingredients.
2. Quantifying chance like finding odds and probability of events.
3. Scaling a diagram for drafting and architectural uses.
4. Finding percent increase or percent decrease for price mark-ups.
5. Discounts on products based on unit rate.

Convex

A convex shape is a shape where all of its parts "point outwards."

In other words, no part of it points inwards.

For example, a full pizza is a convex shape as its full outline (circumference) points outwards.

A convex shape in Geometry is a shape where the line joining every two points of the shape lies completely inside the shape.

Figure 32. Convex.

Source: Image by Wikimedia commons.

Corresponding Angles

The opening and shutting of a lunchbox, solving a Rubik's cube, and never-ending parallel railway tracks are a few everyday examples of corresponding angles. Corresponding angles are the angles that are formed when two parallel lines are intersected by the transversal. These are formed in the matching corners or corresponding corners with the transversal. The corresponding angles definition tells us that when two parallel lines are intersected by a third one, the angles that occupy the same relative position at each intersection are known to be corresponding angles to each other.

Corresponding Angles Theorem

According to the corresponding angles theorem, the statement "If a line intersects two parallel lines, then the corresponding angles in the two intersection regions are congruent" is true either way. Thus, the corresponding angles converse theorem would be, "If the corresponding angles in the two intersection regions are congruent, then the two lines are said to be parallel."

What if a transversal intersects two parallel lines and the pair of corresponding angles are also equal? Then, the two lines intersected by the transversal are said to be parallel. This is the converse of the corresponding angle theorem.

Cubic Graphs

In the mathematical field of graph theory, a cubic graph is a graph in which all vertices have degree three. In other words, a cubic graph is a 3-regular graph. Cubic graphs are also called trivalent graphs.

A bicubic graph is a cubic bipartite graph.

A cubic function is a polynomial of degree three.

e.g., $y = x^3 + 3x^2 - 2x + 5$

Cubic graphs can be drawn by finding the x and y intercepts. Because cubic graphs do not have axes of symmetry the turning points have to be found using calculus.

Commutative Law

The commutative law entails two elements of a set, say a and b, and an operation * that may be performed on any pair of set elements. According to the commutative law

$a * b = b * a$

This law highlights that the elements' order can be reversed. This formal definition is simpler to understand if the set containing a and b is the set of real numbers and the action * is multiplication.

Comparative Costs

The same product may be advertised differently in a supermarket, and customers are eager to get the best deal, or the best value for their money.

Complementary Addition

This is a way for subtracting two numbers by utilizing addition. Prior to the installation of automated cash registers in shops, the shopkeeper would frequently utilize a kind of complimentary addition when providing change to a customer. Assume John paid $17 for a pen and provided the shopkeeper a $50 cash in exchange. The shopkeeper begins counting the change, beginning with the cost of the pen, which is $17. He gives John $3, bringing the total to $20. Then he gives him $30, bringing the total to $50.

Components of a Vector

An example is used to explain the topic of vector components in two dimensions. If R is a vector, it is expressed as R in the notation.

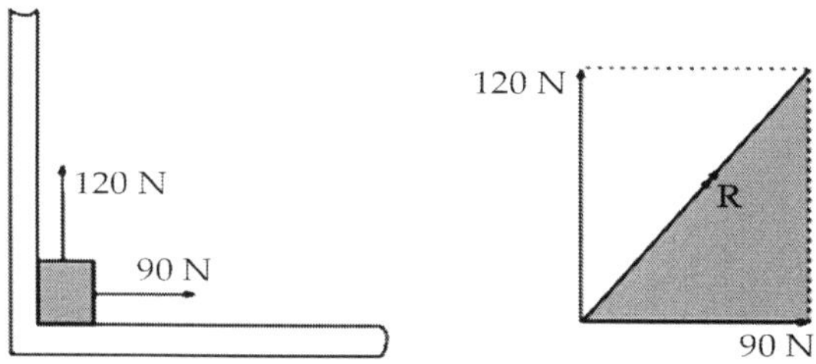

Figure 33. Components of a vector.

Source: Image by Noun Project.

Composite Bar Graphs

A composite bar graph is constructed when two or more bar graphs with comparable information are combined together on the same axis.

Composite Shapes

These are shapes composed of two or more different shapes.

Composite Transformations

These are also known as combined transformations. A composite transformation occurs when a final image is created after two or more transformations in a way that the first image becomes the object for the second transformation. Capital letters are frequently used to denote transformations.

Compound Interest

Compound interest is a business term that refers to the application of percentages to the investment of money. The following is a definition of the terminology used:

1. The principal, represented by P, is the money that an individual invests in the bank.
2. The principal generates interest, symbolized by I, which is money given to the customer by the bank as a reward for utilizing their money.
3. The rate of interest, denoted by R%, is the percentage of the customer's principal that is returned to them.

4. Per annum implies "for each year."
5. The amount, marked by A, is the principal plus the interest.

Concave

Concave is a term used to describe polygons and curves. A concave polygon is one with at least one interior angle greater than 180°. A concave polygon, in other words, has at least one interior angle that is a reflex angle. A reentrant polygon is another name for a concave polygon.

Concentric Circles

When two circles share the same center, they are said to be concentric circles. When a small stone is thrown into the center of a bowl of still water, the waves flow outward in concentric circles from the center of the bowl. The target in the game of darts is formed of concentric circles of wire that are secured to a circular board.

Concurrent

Lines are said to be concurrent when they all pass through the same point. An example can be found in the entry Orthocenter.

Cone

A cone is a three-dimensional structure that is close to a pyramid but has a closed curve at its base. The cone under consideration has a circle for a base and is a right cone, meaning its axis of symmetry is at right angles to its base. Figure a shows the cone's sections labelled. A generator is the straight line having the least distance drawn on the curved surface of the cone from the vertex to a point on the base's circumference.

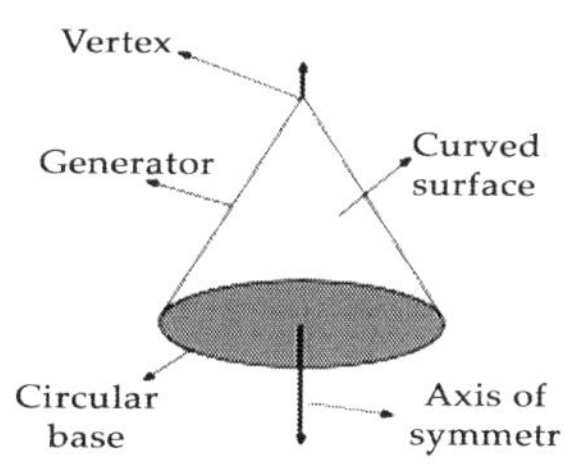

Figure 34. Cone.

Source: Image by Wikimedia Commons.

Congruent Figures

Two figures are congruent if they have the same shape and size and can be laid perfectly on top of each other.

The figures are oppositely congruent if one figure must be turned over to lie exactly on top of the other. the two figures are directly congruent if neither figure needs to be turned over. Figure a shows two pentominoes that are oppositely congruent as one of them must be turned over to lie exactly on top of the other.

The two pentominoes in figure b are directly congruent since one will lie exactly on top of the other without needing to be turned over, but one will need to be turned around. Since the object and image are directly congruent for rotation and translation transformations, they are referred to as direct transformations.

Reflection is an opposite, indirect transformation because the object and image are oppositely congruent. Reflection is a type of indirect transformation. When the object and image have congruent forms, the transformation is known as an isometric transformation. Rotation, translation, and reflection are all isometric. Enlargement is not an isometric because the object and image are not congruent due to their different sizes.

Conic Sections

Conic sections are a collection of curves created when a plane surface cuts a cone, such as a circle, parabola, ellipse, or hyperbola. Each of these curves is discussed one at a time.

Circle- When the flat surface is parallel to the base of the cone, this conic section is obtained.

Parabola- When the plane surface is parallel to one of the cone's generators, this conic section is generated.

Ellipse- When the plane surface is not parallel to one of the generators or the cone's base, this conic section is produced.

Hyperbola- When the flat surface is perpendicular to the bases of two cones, this conic section is created. Since the hyperbola contains two branches, two cones are required.

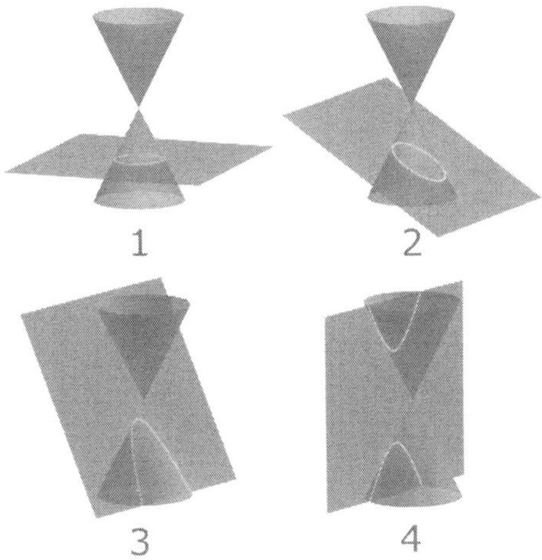

Figure 35. Conic section.

Source: Image by Wikimedia commons.

Constants

Constants are the symbols used to represent integers with fixed values. These symbols could be actual numbers or letters of the alphabet that represent numbers. Constants appear in formulas and can also be found in algebraic words and expressions. A constant is used when the value of a number has not been specified but may be provided later. It is a fixed and unchangeable value for a specific calculation.

Conversion

Conversions are done to convert the units of measurement of a quantity. Temperature, for example, is measured in degrees Celsius (°C) and degrees Fahrenheit (°F). It may be essential to shift from one unit to the other. A formula is required to alter the units, as seen in the following example.

Coplanar

Coplanar points, lines, and polygons are those that lie on the same plane. Lying on the same plane means being on the same level surface. The question of whether or not points, lines, and polygons are coplanar occurs exclusively in three-dimensional models with more than one plane.

Corollary

This is an additional theorem that proceeds a previously proven primary theorem. Consider the angle at the center theorem, which is a circle geometry theorem as an example. The key theorem is as follows: The angle subtended by the arc AB at the center O of a circle is equal to twice the angle subtended by the same arc at any point C on the circumference of the same circle. In terms of symbols, this indicates that Angle AOB = 2 × angle ACB.

This primary theorem's corollary is as follows: Assume that the minor arc AB is extended to form a semicircle. The angle AOB in the center is now 180°, and the angle ACB at the circumference is 90°. The angle at the center theorem's corollary is now: Angle in a semicircle is a right angle.

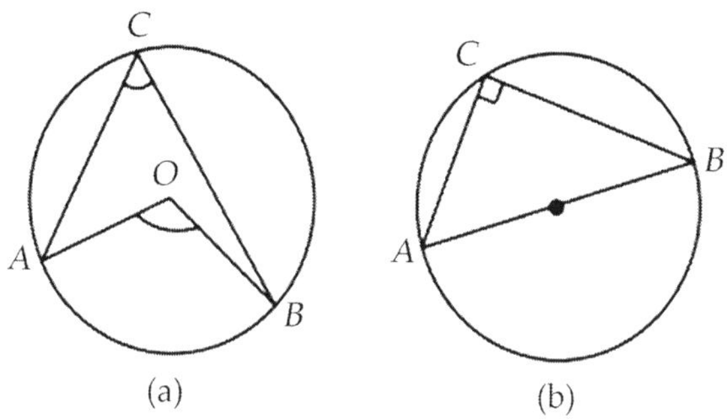

Figure 36. Corollary.

Source: Image by Wikimedia commons.

Cosine Rule

The cosine rule is a collection of trigonometric formulas that connect the three sides of a triangle with the cosine of one of the triangle's angles. It is primarily employed in non-right-angled triangles because easier methods are utilized for right-angled triangles, as detailed in the entry Trigonometry. The sizes of the three angles of the triangle are referred to using capital letters, such as A, B, and C, and the lengths of the sides are referred to using lower case letters, such as a, b, and c. The standard is that side a is the obverse of angle A, and vice versa for sides b and c.

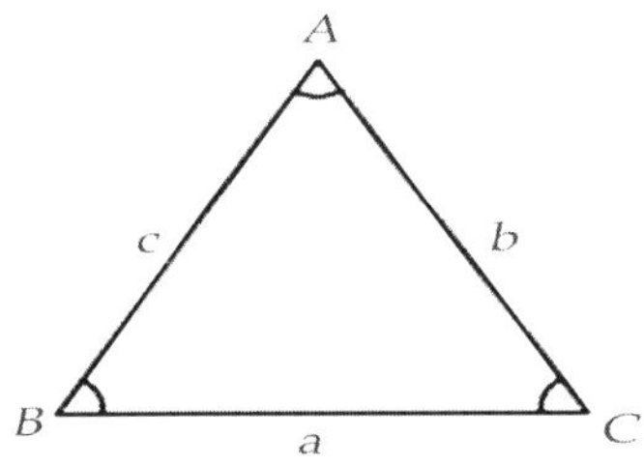

Figure 37. Cosine.

Source: Image by Wikipedia.

The cosine rule asserts that in any triangle, $a^2 = b^2 + c^2 - 2bc \cos A$, and this rule is used to calculate the length of side a.

Cost Price

Amanda owns a sporting goods store. When she purchases a pair of running shoes from a manufacturer to sell in her store, the price at which the shoes are purchased is referred to as the cost price (CP). The selling price is the price at which the shoes are sold to a customer (SP). Her profit or markup is the difference between these two prices. This can be expressed as follows:

Profit = SP – CP

Counterexample

This is a specific example that, if discovered, is used to invalidate a formula or theory that is widely held to be true. A general theory cannot be valid if there is at least one case in which it is not true.

Cross Section

A flat surface formed when a plane cuts through a solid shape. A saw cutting through a block of wood is a fitting analogy. Any angle can be used to make the "saw cut." This is depicted in the figure, where the saw is cutting through a solid block of wood.

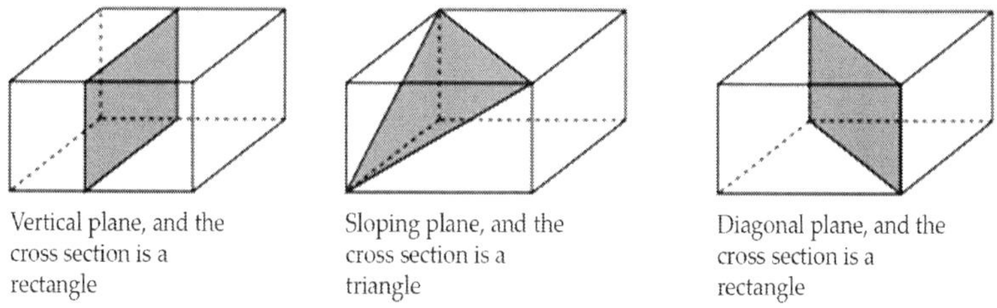

Figure 38. Cross section.

Source: Image by Wikipedia.

Cube (Algebra)

In algebra, a number, quantity, or term is cubed by multiplying it by itself three times. For example, the number 4 can be cubed by calculating $4 \times 4 \times 4 = 64$. 4 cubed equals 64, or that the cube of 4 equals 64. Using exponent notation, the abbreviation for $4 \times 4 \times 4$ is 4^3, thus $4^3 = 64$ may be written.

Cube (Geometry)

One of the five Platonic solids is the cube. It is also known as a regular hexahedron, which refers to a solid with six identical plane faces. Its faces are all squares, the same size, and intersect at right angles. When the length of each side of the cube is x units, the volume of the cube is equal to x 3 cubic units.

Cubic Equations

A cubic equation is one in which 3 is the greatest power to which the variable can be raised.

Cuboid

This is a hexahedron, a solid object with six rectangular faces that intersect at right angles. The hexahedron does not have regular faces; otherwise, it would be a cube with square faces. A rectangular block is another name for a cuboid. The cuboid's dimensions are length, width, and height, indicated by the letters L, W, and H, respectively.

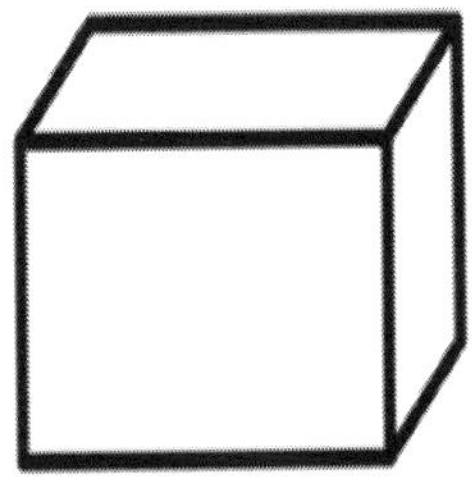

Figure 39. Cuboid.

Source: Image by Noun Project.

Cumulative Frequency Graph

An example is used to explain the cumulative frequency graph. It is particularly useful for determining the median, upper quartile, lower quartile, and interquartile range of a set of continuous data. David's son, William, attends a local high school where all of the pupils are working on a statistics project regarding their heights. The school has 100 students, and their heights are measured and documented in a frequency table. In the table, their heights are recorded in 5 cm class intervals. The class interval "140–" indicates that the heights range from 140 to 145, including 140 but excluding 145. It is important to note that the final cumulative frequency of 100 is always the total frequency.

Currency Conversions

Different countries throughout the world use different currencies, and when travelling abroad, domestic currency must be exchanged for some of theirs in order to buy items there. The exchange rate of one currency for another fluctuates from day to day, and banks have the most recent rate.

Cycle

A cycle is a single repetition of a periodic graph, which is a graph that repeats itself at periodic intervals. A sine curve, whose equation is $y = \sin x$, is an example of a periodic graph. This sine curve's cycle is shown in the figure, and it ranges from $x = 0°$ to $x = 360°$. This section of the curve is repeated every 360°, hence the period of this cyclic curve is 360°. The period of a curve is specified by x values.

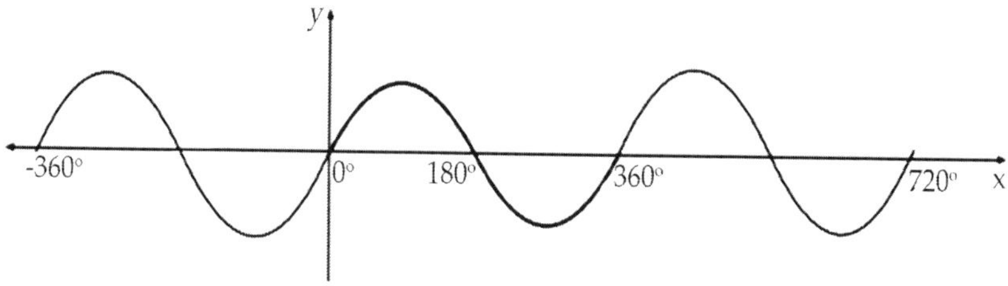

Figure 40. Cycle

Source: Image by Wikimedia Commons

Cycloid

This is the locus drawn by a point on the circumference of a circle when the circle moves along a line without moving. The numerous places of the point on the rolling circle are connected with a dashed line in the figure to show how the cycloid appears. Consider the point to be on the tread of a tyre as it rolls down a flat road.

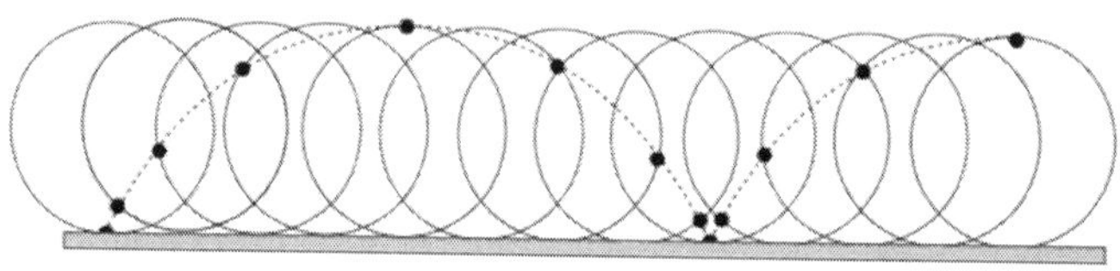

Figure 41. Cycloid.

Source: Image by Wikimedia Commons.

Cylinder

A cylinder is a solid that is made up of three surfaces. The cylinder discussed here is composed of two parallel-planed circles connected to a tube constructed by rolling up a rectangle. The cylinder studied here is erect, with its axis of symmetry at right angles to its base; this is known as a right cylinder. A cylinder is a circular-sectioned prism. The cylinder's net is made up of a rectangle that rolls up to form a tube and two circles. Each circle has a point where it connects to the rectangle. Hence,

Length of the rectangle = circumference of the circle

Length = $2\pi R$

D

Data

The data refer to the collection of quantities or information obtained from a sample of a population during a survey or experiment. Statisticians gather, display, and analyze drawing meaningful conclusions and make predictions.

The heights of the students at Washington College, for example, are a set of data. Another source of data is the nationalities of the teachers at Washington College.

1. Qualitative data are descriptive data, such as eye color, nationality, gender, job categories, political beliefs, and so on.
2. Quantitative data are data on numbers, such as the number of employees in a plant, the number of children in a household, the temperature of a liquid, and so on. Such data can be sorted according to size.
3. Nominal data are qualitative data that cannot be classified in any particular order. For example, car colors cannot be ranked because one color cannot be considered higher or lower than another.
4. Ordinal data are qualitative data that can be ranked in some way. A person's attitude toward television, for example, can be divided into three categories: likes it, finds it OK, or dislikes it.
5. Discrete data are quantitative data that can't be subdivided and are obtained by counting. The number of individuals in a family, for example, is a counting number that cannot be divided in half.
6. Continuous data are quantitative data gathered through measurement, with each item of measurement having an endless number of possible values limited by only the restrictions of the measuring apparatus, such as the speed of cars passing a given milestone on a highway.

Decagon

A 10-sided polygon is known as a decagon. Figure a depicts several decagons, the last of which is a regular decagon. The regular decagon has ten equal-length sides and ten angles that add up to 144°. It has ten axes of symmetry and a rotational symmetry order of ten. The regular decagon is composed of 10 congruent isosceles triangles with angles of 36°, 72°, and 72°. Tessellation will not occur with a regular decagon.

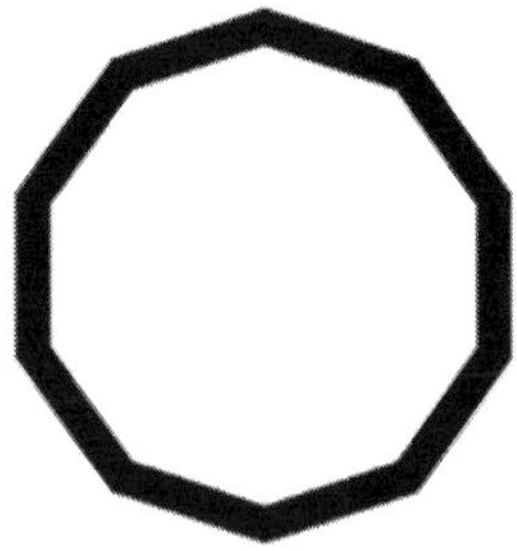

Figure 42. Decagon.

Source: Image by Noun Project.

Decimal

A decimal fraction is another name for a decimal. A decimal is the same as a proper fraction with a denominator of 10, 100, 1000, or higher powers of 10, except it is written without a denominator and with a decimal point. A decimal number can be negative.

Decimal System

The decimal system, also known as the denary system, is a numerical system based on ten and powers of ten. It developed in India and was brought to Europe by the Arabs. The decimal system is used for counting, money, weights and measures, and so on. To write numbers, the decimal number system employs the digits 0, 1, 2, 3, 4, 5, 6, 7, 8, and 9. The value of each of these digits 0–9 is determined by its position in the decimal number. In the decimal number 7325, for example, the 7 represents the value $7\times10^3 = 7000$, the 3 represents the value $3\times10^2 = 300$, the 2 represents the value $2\times10^1 = 20$, and the 5 represents the value $5\times100 = 5$.

Decile

A decile is a quantitative method of splitting up a set of ranked data into 10 equally large subsections. This type of data ranking is performed as part of many academic and statistical studies in the finance and economics fields. The data may be ranked from largest to smallest values, or vice versa.

A decile, which has 10 categorical buckets may be contrasted with percentiles that have 100, quartiles that have four, or quintiles that have five.

In descriptive statistics, a decile is used to categorize large data sets from the highest to lowest values, or vice versa. Like the quartile and the percentile, a decile is a form of a quantile that divides a set of observations into samples that are easier to analyze and measure.

While quartiles are three data points that divide an observation into four equal groups or quarters, a decile consists of nine data points that divide a data set into 10 equal parts. When an analyst or statistician ranks data and then splits them into deciles, they do so in an attempt to discover the largest and smallest values by a given metric.

For example, by splitting the entire S&P 500 Index into deciles (50 firms in each decile) using the P/E multiple, the analyst will discover the companies with the highest and lowest P/E valuations in the index.

A decile is usually used to assign decile ranks to a data set. A decile rank arranges the data in order from lowest to highest and is done on a scale of one to 10 where each successive number corresponds to an increase of 10% points. In other words, there are nine decile points. The 1st decile, or D1, is the point that has 10% of the observations below it, D2 has 20% of the observations below it, D3 has 30% of the observations falling below it, and so on.

Decimal Places

If a number has a decimal point , then the first digit to the right of the decimal point indicates the number of tenths.

For example, the decimal 0.30.3 is the same as the fraction 3/10.

The second digit to the right of the decimal point indicates the number of hundredths.

One good way to visualize decimals is by using base 10 blocks. For instance, suppose a large square represents one whole. If the square is cut into 10 strips of equal size, then each of these represents one tenth or 0.1. Each strip can be cut into ten smaller squares to represent hundredths.

Denary Numbers

A denary number is a number in the base 10, or decimal, system. Most of the numbers used internationally are denary numbers, with a few exceptions possible in specific fields such as computer science.

Numerals

The base of a numeral system refers to the number of numerals that are used for expressing a number in that system. Denary numbers use 10 numerals. These are "0," "1," "2," "3," "4," "5," "6," "7," "8," and "9."

Place Value

Digits in denary numbers have a place value given by their position within the number. If there is no decimal point, the rightmost digit is in the "ones" place, which has the value of the digit times 10^0 (10 raised to the zeroth power, or 1).

Fractional Values

Digits to the right of the decimal point in denary numbers show portions of a whole. The value of each digit is determined by its place. The value of the first digit to the right of the decimal point is the digit multiplied by $10^{(-1)}$, or 1/10. Each digit to the right of the decimal point has a value of the digit multiplied by $10^{(-n)}$, where "n" stands for the number of places to the right of the decimal point.

Identification

Base 10 is the common numeral system used internationally. Unless otherwise specified, a number can be assumed to be denary.

Other Base Systems

Besides base 10 numeral system, computer scientists may use base 2 (binary), base 8 (octal) and base 16 (hexadecimal) numeral systems.

Diameter

The diameter of a circle is any straight line segment that passes through the center of the circle and whose endpoints lie on the circumference of a circle. The diameter is also known as the longest chord of the circle.

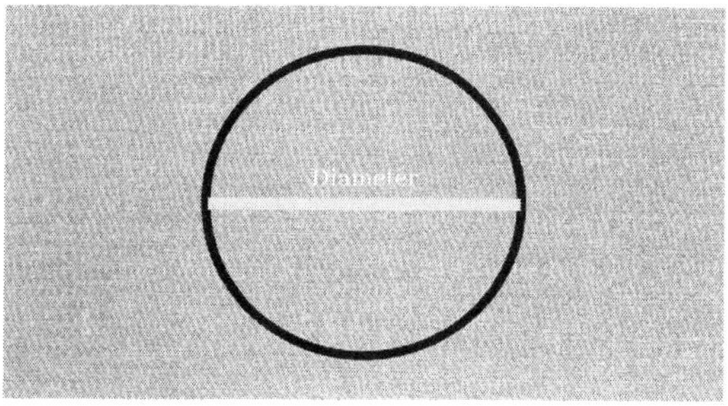

Figure 43. Diameter.

Source: Image by Pixabay.

Diameter Definition

The diameter is defined as twice the length of the radius of a circle. The radius is measured from the center of a circle to one endpoint of the circle, whereas, the distance of diameter is measured from one end of the circle to a point on the other end of the circle, passing through the center. The diameter is denoted by the letter D. There are infinite points on the circumference of a circle, this means that a circle has an infinite number of diameters, and each diameter of the circle is of equal length.

Discount

Discount is the reduction in the price of goods or services offered by shopkeepers at the marked price. This percentage of the rebate is usually offered to increase the sales or clear the old stock of goods. The List price or Marked price is the price of an article as declared by the seller or the manufacturer, without any reduction in price. Selling price is the actual price at which an article is sold after any reduction or discounts in the list price. "Off," "Reduction" are some common terms used to describe discounts. It should be noted that discount is always calculated on the Marked price (List price) of the article. The formula to calculate discount is:

Discount = List Price – Selling Price

Discount (%) = (Discount/List Price) × 100

Discrete Distribution

A discrete distribution is a probability distribution that depicts the occurrence of discrete (individually countable) outcomes, such as 1, 2, 3… or zero vs. one. The binomial distribution, for example, is a discrete distribution that evaluates the probability of a "yes" or "no" outcome occurring over a given number of trials, given the event's probability in each trial—such as flipping a coin one hundred times and having the outcome be "heads."

Statistical distributions can be either discrete or continuous. A continuous distribution is built from outcomes that fall on a continuum, such as all numbers greater than 0 (which would include numbers whose decimals continue indefinitely, such as pi = 3.14159265…). Overall, the concepts of discrete and continuous probability distributions and the random variables they describe are the underpinnings of probability theory and statistical analysis.

Distribution is a statistical concept used in data research. Those seeking to identify the outcomes and probabilities of a particular study will chart measurable data points from a data set, resulting in a probability distribution diagram. There are many types of probability distribution diagram shapes that can result from a distribution study, such as the normal distribution ("bell curve").

Statisticians can identify the development of either a discrete or continuous distribution by the nature of the outcomes to be measured. Unlike the normal distribution, which is continuous and accounts for any possible outcome along the number line, a discrete distribution is constructed from data that can only follow a finite or discrete set of outcomes.

Discrete distributions thus represent data that has a countable number of outcomes, which means that the potential outcomes can be put into a list. The list may be finite or infinite. For example, when studying the probability distribution of a die with six numbered sides the list is {1, 2, 3, 4, 5, 6}. A binomial distribution has a finite set of just two possible outcomes: zero or one—for instance, flipping a coin gives you the list {Heads, Tails}. The Poisson distribution is a discrete distribution that counts the frequency of occurrences as integers, whose list {0, 1, 2, …} can be infinite.

Distance

Distance is a numerical description of how far apart two objects are. There are many different ways to determine the distance between two objects.

In addition, there are just as many tools that you can use. Mathematically, if you want to determine the distance between two points on a coordinate plane, you use the distance formula.

$d = \sqrt{(x2 - x1)^2 + (y2 - y1)^2}$

When you know the coordinates of the two points that you're trying to find the distance between, just substitute them into the equation. It doesn't really matter which point is $(x1, y1)$ or which one is $(x2, y2)$ – just so long as you keep them together. Whichever set you use for 1, use it for both $x1$ and $y1$, and whichever set you use as 2, use both $x2$ and $y2$ from that set.

Distributive Law

Distributive law, in mathematics, the law relating the operations of multiplication and addition, stated symbolically, $a(b + c) = ab + ac$; that is, the monomial factor a is distributed, or separately applied, to each term of the binomial factor $b + c$, resulting in the product $ab + ac$. From this law it is easy to show that the result of first adding several numbers and then multiplying the sum by some number is the same as first multiplying each separately by the number and then adding the products.

Divisor

A divisor is a number that divides another number. Without a divisor, you cannot divide numbers. In a division, there are four important terms, they are dividend, divisor, quotient, and remainder. The division is the process of making equal groups.

The total number of objects to be divided takes the name of a 'Dividend' and the total number of equal groups to be formed is the 'Divisor.' If there are any objects left out without forming a group they are termed as 'Remainder.' A number that divides another number with or without leaving a remainder is called a divisor. Divisor takes the dividend and divides it into equal groups. The number being divided in a division problem is called a dividend and the number that the dividend is divided by is called the divisor.

Domain

1. The set of all possible values which qualify as inputs to a function is known as the domain of the function, or it can also be defined as the entire set of values possible for independent variables.

2. The domain can be found in – the denominator of the fraction is not equal to zero and the digit under the square root bracket is positive. (In case of a function with fraction values).

Decimal Number

A real number which expresses fractions on the base 10 standard numbering system using place value, e.g., $^{37}/_{100} = 0.37$.

A decimal number system is used to express the whole number and fraction together. Here, we will separate the whole number from the fraction by inserting a.”,” which is called a decimal point.

Figure 44. Decimal number.

Source: Image by Pixabay.

Deductive Reasoning or Logic

A type of reasoning where the truth of a conclusion necessarily follows from, or is a logical consequence of, the truth of the premises (as opposed to inductive reasoning)

Derivative

A measure of how a function or curve changes as its input changes, i.e., the best linear approximation of the function at a particular input value, as represented by the slope of the tangent line to the graph of the function at that point, found by the operation of differentiation.

Derivative, in mathematics, the rate of change of a function with respect to a variable. Derivatives are fundamental to the solution of problems in calculus and differential equations. In general, scientists observe changing systems (dynamical systems) to obtain the rate of change of some variable of interest, incorporate this information into some differential equation, and use integration techniques to obtain a function that can be used to predict the behavior of the original system under diverse conditions.

Geometrically, the derivative of a function can be interpreted as the slope of the graph of the function or, more precisely, as the slope of the tangent line at a point. Its calculation, in fact, derives from the slope formula for a straight line, except that a limiting process must be used for curves. The slope is often expressed as the "rise" over the "run," or, in Cartesian terms, the ratio of the change in y to the change in x.

Descriptive Geometry

A method of representing three-dimensional objects by projections on the two-dimensional plane using a specific set of procedures

Differential Equation

An equation that expresses a relationship between a function and its derivative, the solution of which is not a single value but a function (has many applications in engineering, physics economics, etc.).

Differential Geometry

A field of mathematics that uses the methods of differential and integral calculus (as well as linear and multilinear algebra) to study the geometry of curves and surfaces.

Differential geometry, branch of mathematics that studies the geometry of curves, surfaces, and manifolds (the higher-dimensional analogs of surfaces). The discipline owes its name to its use of ideas and techniques from differential calculus, though the modern subject often uses algebraic and purely geometric techniques instead.

While curves had been studied since antiquity, the discovery of calculus in the 17th century opened up the study of more complicated plane curves—such as those produced by the French mathematician René Descartes (1596–1650) with his "compass." In particular, integral calculus led to general solutions of the ancient problems of finding the arc length of plane curves and the area of plane figures. This in turn opened the stage to the investigation of curves and surfaces in space—an investigation that was the start of differential geometry.

Some of the fundamental ideas of differential geometry can be illustrated by the strake, a spiraling strip often designed by engineers to give structural support to large metal cylinders such as smokestacks. A strake can be formed by cutting an annular strip (the region between two concentric circles) from a flat sheet of steel and then bending it into a helix that spirals around the cylinder. Differential geometry supplies the solution to this problem by defining a precise measurement for the curvature of a curve; then r can be adjusted until the curvature of the inside edge of the annulus matches the curvature of the helix.

Differentiation

The operation in calculus (inverse to the operation of integration) of finding the derivative of a function or equation.

In calculus, differentiation is one of the two important concepts apart from integration. Differentiation is a method of finding the derivative of a function. Differentiation is a process, in Math's, where we find the instantaneous rate of change in function based on one of its variables. The most common example is the rate change of displacement with respect to time, called velocity. The opposite of finding a derivative is anti-differentiation.

If x is a variable and y is another variable, then the rate of change of x with respect to y is given by dy/dx. This is the general expression of derivative of a function and is represented as f'(x) = dy/dx, where y = f(x) is any function.

Diophantine Equation

A polynomial equation with integer coefficients that also allows the variables and solutions to be integers only.

Diophantine equation involving only sums, products, and powers in which all the constants are integers and the only solutions of interest are integers. For example, $3x + 7y = 1$ or $x^2 - y^2 = z^3$, where x, y, and z are integers. Named in honor of the 3rd-century Greek mathematician Diophantus of Alexandria, these equations were first systematically solved by Hindu mathematicians beginning with Aryabhata (c. 476–550).

Diophantine equations fall into three classes: those with no solutions, those with only finitely many solutions, and those with infinitely many solutions. For example, the equation $6x - 9y = 29$ has no solutions, but the equation $6x - 9y = 30$, which upon division by 3 reduces to $2x - 3y = 10$, has infinitely many. For example, $x = 20$, $y = 10$ is a solution, and so is $x = 20 + 3t$, $y = 10 + 2t$ for every integer t, positive, negative, or zero. This is called a one-parameter family of solutions, with t being the arbitrary parameter.

Congruence methods provide a useful tool in determining the number of solutions to a Diophantine equation. Applied to the simplest Diophantine equation, $ax + by = c$, where a, b, and c are nonzero integers, these methods show that the equation has either no solutions or infinitely many, according to whether the greatest common divisor (GCD) of a and b divides c: if not, there are no solutions; if it does, there are infinitely many solutions, and they form a one-parameter family of solutions.

Distributive Property

Property whereby summing two numbers and then multiplying by another number yields the same value as multiplying both values by the other value and then adding them together, e.g., $a(b + c) = ab + ac$

Decomposition

The term "decomposition" refers to the process of breaking down a number or quantity into smaller components. The number 456 is made up of 400 + 50 + 6, which can be rearranged to form 400 + 40 + 16. This latter form of the number has the same value as the first, but has been decomposed. During subtraction, it can be useful to use the decomposition of one of the numbers.

Decreasing Function

The concept of a strictly decreasing function is shown with the use of a specific function, $y = x^2$, which is defined for all real integers. Assume A in figure a is the point with coordinates (–3, 9) and B is the point with coordinates (–2, 4). The x values rise in size from- 3 to –2 across the interval from A to B, however the y values decrease in size from 9 to 4. As x increases, y decreases, and the function $y = x^2$ is said to be strictly decreasing from A to B. The gradient of the line segment AB is negative, implying that the function is strictly decreasing over this interval.

Degree

The word degree, as it is commonly used in mathematics, has three separate meanings.

1. A degree is one of the units used to calculate the size of an angle, which is a turning amount. Radians and gradians are two other units. A degree is a very little amount of turning that is 1/360 of a full turn or complete turn in size. A full turn has 360 degrees; written as 360° in symbols. A protractor is used to measure angles in degrees. A person in the army would quickly learn that a half turn is the same as an "about turn" and is a 180-degree rotation. The Babylonians established this division of a full turn into 360 equal parts, known as degrees, over 3000 years ago and it is being used today. Since it encompasses so many variables, 360 is a handy figure to employ.
2. A degree is a unit of measurement used to measure the temperature of something with a thermometer. Degrees Celsius and degrees Fahrenheit are two units that can be used.
3. The degree of a polynomial is the variable's highest power. The polynomial $3x^4 - 2x^2 + 1$ is of degree 4, for example, because the highest power of the variable x is 4. A polynomial has no negative powers.

Delta

The fourth letter of the Greek alphabet is; the lower – case letter is δ and the capital letter is Δ. In mathematics, the capital letter is typically used as the discriminant of a quadratic equation or as a very small but finite increase of a variable. For instance, Δx is a minor increase in x.

Figure 45. Delta.

Source: Image by Noun Project.

Denominator

A fraction is composed of 2 numbers placed one on top of the other. The top number is known as the fraction's numerator, while the lower number is known as the fraction's denominator. The numerator is often referred to as the dividend, while the denominator is often referred to as the divisor. For example, in the fraction 2/3 the numerator is 2 and the denominator is 3. The improper fraction 19/7, with a numerator of 19 and a denominator of 7, must first be expressed as the mixed number 2 5/7. The algebraic fraction 3ab/4x 2 has 3ab as the numerator and 4x 2 as the denominator.

Dependent Variable

Assume William has $30 right now and chooses to save $5 each week. Suppose he has saved $y after x weeks. y = 5x + 30 is the equation that links the variables x and y. The quantity of money saved (y) in this equation is dependent on the number of weeks (x) he has been saving; hence y is referred to as the dependent variable. The independent variable is the number of weeks x. When the equation is graphed, the independent variable x is placed on the horizontal axis and the dependent variable on the vertical axis.

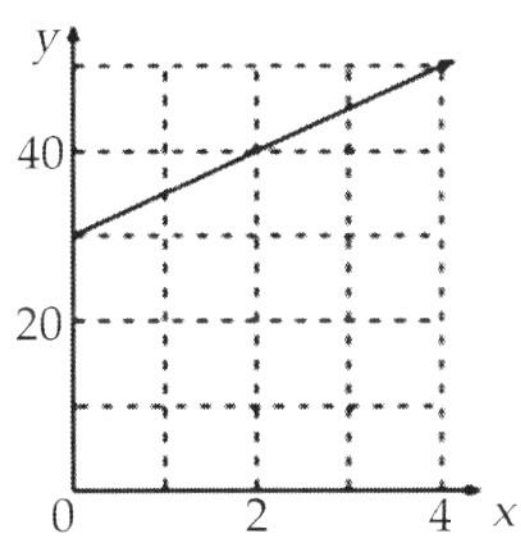

Figure 46. Dependent variable.

Source: Image by Pixabay.

Depth

This is the distance between the top and bottom of an object, while the height is the distance from the bottom and the top. As a result, the depth and height are the same distance when viewed from different perspectives. When viewed from the surface, the depth of a hole or the depth of a swimming pool is considered, but when viewed from ground level, the height of a tower or a power pole is seen.

Deviation

This is the distinction between one quantity and another. One of the quantities is typically a fixed or standard quantity. The intelligence quotient, also known as IQ, is one of the measures of intellect. The IQ of a super brain is 140. If Walton has an IQ of 128, his brain's deviation from a super brain equals

Deviation = 140 – 128 = 12

Diagonal

A polygon's diagonal is a line that connects any two vertices that are not adjacent to each other on the polygon. For example, figure a shows pentagon ABCDE, where the line AD is a diagonal, as is BD. Since A and E are adjacent vertices on the pentagon, the line AE is not a diagonal. The pentagon has five diagonals, which are shown in the figure as dashed lines.

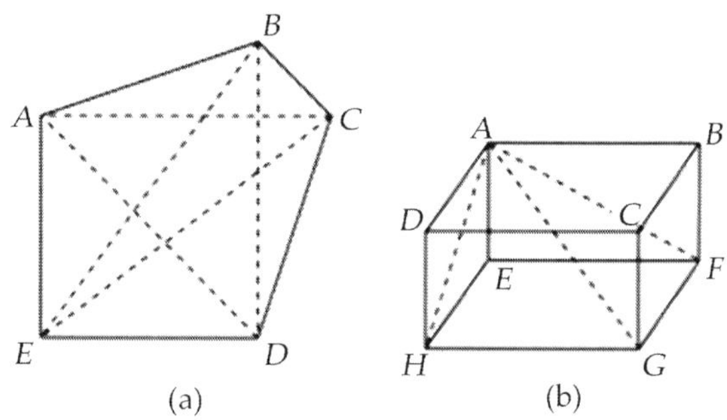

Figure 47. Diagonal.

Source: Image by Wikimedia commons.

Dice

The plural of the word die is dice. A die is a tiny cube, typically made out of plastic, with dots representing the numbers 1, 2, 3, 4, 5, and 6. Since a dice is employed in many board games and games of chance, it raises interesting probability problems. The numbers on its faces are placed in such a way that 1 is opposite 6, 2 is opposite 5, and 3 is opposite 4. They are placed in such a way that the numerals on opposite sides total up to 7.

Figure 48. Dice.

Source: Image by Flickr.

Difference Tables

The difference tables can be defined as the tables which are used to find a formula (or general term) for a sequence of numbers. It is possible to examine only two types of difference tables. The very first is related to a linear formula and the second is related to a quadratic formula.

Digit

A digit can be defined as any one of the numerals 0, 1, 2, 3, 4, 5, 6, 7, 8, 9. Let's take an example, the number 35 is made up of two digits and those are 3 and 5.

Dihedral Angle

A dihedral angle can be referred to as the angle that is between two planes which tend to intersect in a straight line. Let's suppose that there are two planes which intersect in a straight-line AB (consider the figure) and point P is any point on the line AB. Then, a straight line PX is drawn in one plane in such a way that it makes a right angle with the line AB. In the similar manner, a line PY is drawn in the other plane so that it makes a right angle along with the line AB. The angle XPY will be the dihedral angle.

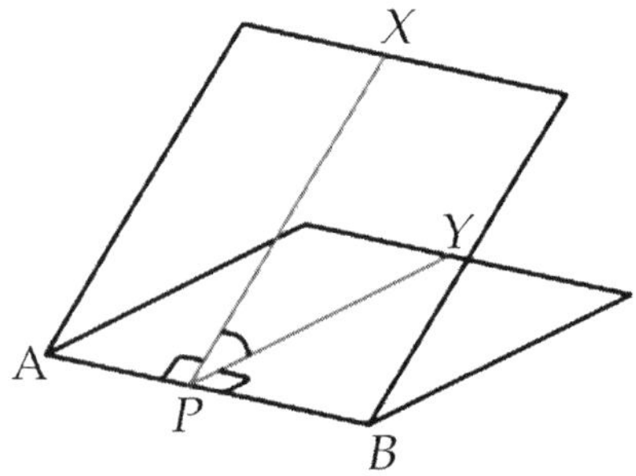

Figure 49. Dihedral angle.

Source: Wikimedia Commons.

Dilation

Dilation can say to be another word or another name for enlargement.

Discontinuous

A function is said to be discontinuous if there exists a case when its graph is broken into two or more parts that are not connected. If a graph of a function is drawn, at a point of discontinuity, it will require taking the pencil off the paper to continue the drawing of the graph.

Displacement

The displacement can be adequately described and there will be a better to say what distance is. This is because of the fact that displacement and distance are similar. Distance is referred to as the measure of the change in position of an object or point and moreover, can take place in any direction. It is generally the length of the straight line which can be drawn from the starting point (or position) to the final point (position). The units of distance are millimeters (mm), meters (m), centimeters (cm), and kilometers (km). Displacement can be defined as the shift from the starting position of an object to its final position, but it is important to have the distance and the direction. Its units are the similar as the units of distance, given with the fact that the direction is also stated.

Distribution

Distribution is a term that is widely used in statistics. When the data is collected, that may be a set of observations or a set of measurements, record the frequency of each item of data, and arrange the data in some readable form, it is referred to as a distribution. The distribution of the data can be displayed graphically that is there can be graphs drawn for the same.

Dividend

The dividend is referred to as the number that is being divided when one number is divided by another. Let's take an example to clear the concept. Joanne has 7 Easter eggs and she has 3 children. She has to divide 7 Easter eggs by 3 children in order to get an answer of 2 eggs for each child and 1 egg left over. Considering this example, it can be said that the dividend is 7, the divisor is 3, the quotient is 2, and 1 is the remainder. When 7 is divided by 3 to get an answer of 2 and also, leave a remainder 1. This can be written as a division identity:

$$7 = 3 \times 2 + 1$$

In other words, the division identity is Dividend = Divisor × Quotient + Remainder.

If in case, the remainder is zero, then it can be said that the dividend is exactly divisible by the divisor. This would be the case if in case, Joanne had 6 Easter eggs rather than 7, and also, 6 is divisible by 3, because the remainder is zero.

Divisibility Tests

The divisibility tests are said to be the short cuts used in order to see if one counting number is exactly divisible by another counting number and that too without actually doing the division. These are the tests which tend to provide a check to see if one number is a factor or multiple of another number and this is also done without actually doing the division process.

The divisibility tests were more useful some years ago prior to the calculators which were freely available, but still, they tend to provide a quick test for divisibility.

Test 1. A number is divisible by 10 if in case the number ends in 0.

Test 2. A number is divisible by 5 if in case the number ends in 0 or ends in 5.

Test 3. A number is divisible by 3 if in case the sum of its digits is divisible by 3.

Test 4. A number is divisible by 2 if in case the number ends in 0 or 2, 4, 6, 8. This in turn means that it is an even number.

Test 5. A number is divisible by 4 if in case the number, which is formed by the last two digits, is divisible by 4.

Test 6. A number is divisible by 8 if in case the number formed by the last three digits is divisible by 8.

Test 7. A number is divisible by 9 if in case the sum of its digits is divisible by 9.

Test 8. A number is divisible by 6 if in case the last digit is even and also, the sum of its digits is divisible by 3.

Test 9. A number is divisible by 11 if in case the difference as well as the sum alternately of its digits, that starts with the last digit and works in order to make a number which is either 0 or a number divisible by 11.

Dodecagon

A dodecagon can be defined as a polygon which has 12 sides. In order to find the angle sum of a dodecagon, it is important to look under the entry Polygon. In a regular dodecagon, each interior angle is equal to 150°. The dodecagon has 12 axes of symmetry, and moreover, its order of rotational symmetry is also 12. The regular dodecagon tessellates having two equilateral triangles and one square.

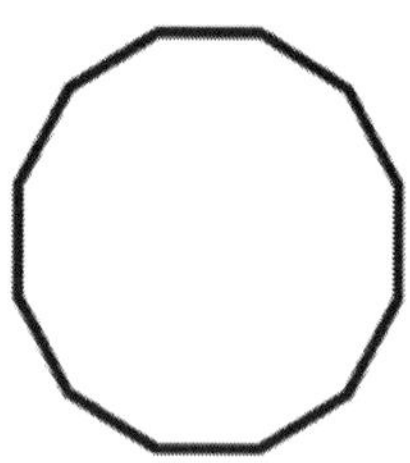

Figure 50. Dodecagon.

Source: Image by Noun Project.

Dodecahedron

The dodecahedron can be referred to as a regular polyhedron (solid) which has 12 congruent faces and also, each and every face is a regular pentagon. The regular dodecahedron is said to be one of the Platonic solids.

Dot Plot

The dot plot is considered to be a graph that is used in statistics and also, it consists of vertical lines of dots in order to show the frequencies rather than columns, as in the bar graph.

Duodecimal

Duodecimal is sometimes referred to as duodenary. The duodecimal is referred to as a number system that is based on 12 and powers of 12. It further uses 12 digits for counting, while the decimal (denary) system of counting uses 10 digits, due to the fact that it is based on 10. The digits which are used in base 12 are {0, 1, 2, 3, 4, 5, 6, 7, 8, 9, t, e}. Here, t is ten and e is eleven. The special symbols for ten and eleven are required and the reason being the fact that the units column goes up to eleven, and 10 and 11 cannot be used, because they occupy two columns.

E

Edge

An edge is referred to as a straight line where two planes of a three-dimensional that is 3D polyhedron meet. A plane of a solid is considered to be a flat surface, and a vertex is a point where three or more planes tend to meet. Sometimes, the side of a polyhedron is referred to as an edge. The polyhedron is a square-based pyramid. It contains eight edges, five planes and five vertices. There exists a formula which connects the number of edges (E), the number of faces (F), and the number of vertices (V) of three-dimensional polyhedra. This is the formula that is referred to as the Euler's formula. This formula is named after the man who discovered it. The formula is

$$F + V = E + 2$$

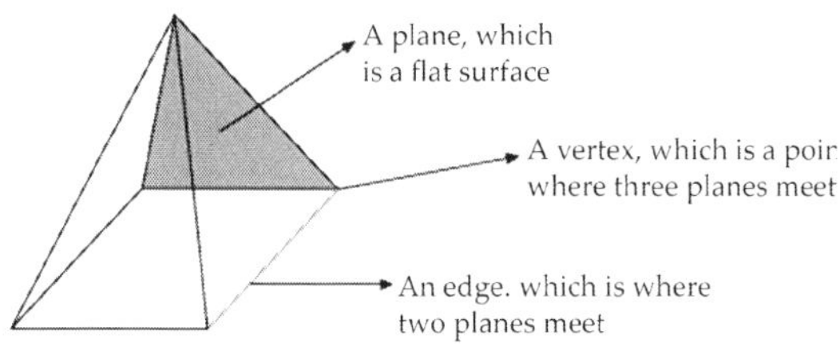

Elevation

An elevation is considered as a view that is drawn accurately to scale, from the front, side or even back of an object.

Elimination

The elimination is referred to as a method of solving simultaneous equations.

Ellipse

An ellipse looks similar to a "flattened circle." There is a common name for an ellipse and that is an oval, but obviously, this is not a word which is used in mathematics. An ellipse contains two axes of symmetry, and also, the order of rotational symmetry is two. The ellipse is considered to be important in science and mathematics due to the fact that the planets move in elliptical orbits around the sun.

There is one method of drawing an ellipse and that is to get a length of string and join its ends together in order to form a loop. For this, place a piece of paper on a drawing board and then, insert in the paper two thumb tacks placed sufficiently far apart so that when the string is looped around them, it is slack. After that, place a pencil in the loop and with it pull the string taut. Now, move the pencil, while keeping the string taut. With this, the shape traced out by the pencil will be an ellipse.

Figure 51. Ellipse (conic).

Source: Image by Wikimedia commons.

Endecagon

The Endecagon is said to be an 11-sided polygon, which is also referred to as a hendecagon. The angle sum of the Endecagon is provided under the entry Polygon.

Enlargement

The enlargement is sometimes referred to as dilation or magnification. An enlargement is considered to be one of the geometrical transformations which tend to occur in everyday life at the movie theater when the film is projected, and further, enlarged onto a screen by passing light rays through the film.

Considering this example, it can be said that the source of light in the projector is the center of enlargement, the film is the object, the picture on the screen is the image and furthermore, the number of times the image is bigger in comparison to the object is the scale factor (k) of the enlargement. In this example, the rays of light correspond to rays that can be drawn with a pencil.

Considering the mathematics when the word enlargement is used, then this do not necessarily mean the image is larger in comparison to the object. It can be said that for an enlargement of one – half, the dimensions of the image will be only half those of the object. In the work which follows it is assumed that the object shape as well as the image shape are in the same plane.

Equations

The equations can be formed for solving some everyday problems. It is the process that is explained under the entry Abstract. While solving an equation, only use one equals sign per line of working, and moreover, keep the equals signs in a straight line under each other.

Equiangular

Taking mathematics into account, the prefix *'equi'* means equal. Thus, an equiangular polygon means a polygon having all its angles equal in size. If in case, a polygon is equiangular and also, all its sides are equal in length, it is referred to as a regular polygon. An equiangular triangle is considered as a triangle having all its angles equal in size. When, the three angles add up to 180°, each and every angle is equal to $180° \div 3 = 60°$. Usually, such a triangle is called equilateral.

Equidistant

If in case, two or more distances are equidistant, then they are equal.

Equilateral Triangle

The equilateral triangle is said to be a triangle having all three of its angles equal in size to 60° and all three of its sides equal in length. An equilateral triangle contains three axes of symmetry and the order of rotational symmetry is three.

Equivalent Equations

The equivalent equations are the steps which are used in solving equations.

Equivalent Expressions

Considering the algebra, two expressions are equivalent if in case, they contain the similar information, but expressed in various forms. The equivalent expressions are expressions which are equal to each other. Let's take an example, $x + 2x$ and $3x - x$ are equivalent expressions. $2x(x + 3)$ is an equivalent expression to $2x\ 2 + 6x$, and further, is obtained by expanding the brackets.

Equivalent Fractions

The fractions are equivalent when in case, they have the similar value, but are written in a number of different ways. A convenient way in order to introduce equivalent fractions is to use areas. Let's suppose a rectangle which has a length of 6 units and a width of 2 units, and furthermore, is divided up in three different ways.

In figure a, the rectangle is divided into 3 equal regions and among them 1 is shaded. This further means that 1/3 of the whole rectangle is shaded. In figure b, the rectangle is divided into 6 equal regions and 2 of them are shaded. In such cases 2/6 of the whole rectangle is shaded. In figure c, in the situation the same rectangle is divided into 12 equal regions and furthermore, 4 of them are shaded. Then, now there is 4/12 of the whole rectangle shaded.

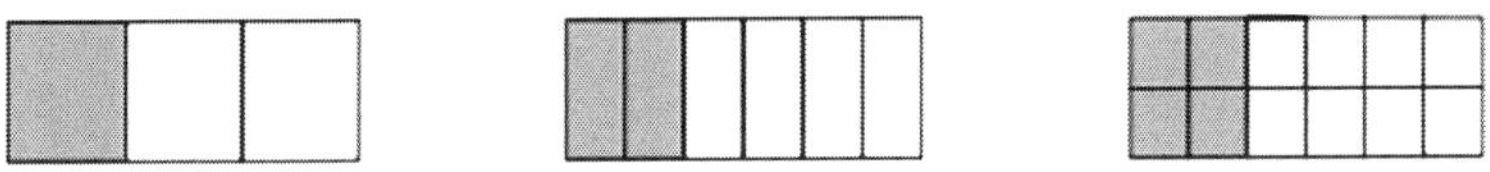

(a) (b) (c)

Eratosthenes' Sieve

Eratosthenes' sieve is considered as an algorithm in order to find prime numbers less than a certain number.

Error

Considering the field of mathematics, the error does not necessarily mean that a mistake has been made, but moreover, that the talk can be about a quantity which is not expressed exactly. Such kind of error may arise because of estimation or due to the fact that it is not possible to make measurements with 100% accuracy.

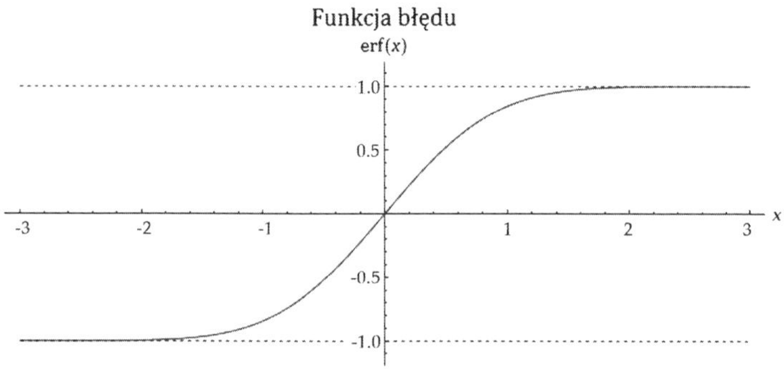

Figure 52. Error (function).

Source: Image by Wikimedia Commons.

Estimation

When estimation is done then calculation takes place for an approximate value for the size of a specific quantity. There are a number of different reasons why the estimation of the size of a quantity rather than attempting in order to find its exact value.

Euler's Formula

Euler's formula can be defined as a formula which was discovered by Leonhard Euler (1707 – 1783) which relates the number of vertices V, edges E and faces F of any simple closed three – dimensional polyhedron. Euler's formula can be written as:

$$F + V = E + 2$$

There is one formula similar to this and this relates the number of nodes N, arcs A and regions R of a network which is drawn in one plane:

$$N + R = A + 2$$

Evaluate

Evaluating an expression means finding the numerical value that is the number value, of the expression.

Even

A number can be referred to as even if it is divisible by 2 having zero remainder. Let's take an example, 6 is divisible by 2 having zero remainder, therefore, 6 is an even number. In the similar manner, −4 is an even number, due to the fact that it is divisible by 2 having zero remainder. The set of even numbers is

{..., −6, −4, −2, 0, 2, 4, 6,...}

Even numbers are generally multiples of 2, and moreover, form an infinite set. The formula for even numbers is **E = 2n**, and the set of even numbers is further obtained and this is done by substituting each of the integer numbers **I = {..., −3, −2, −1, 0, 1, 2, 3,...}** in turn for n. A number is referred to as odd if it is one more or one less in comparison to an even number. The set of odd numbers is

{..., −5, −3, −1, 1, 3, 5,...}

The odd numbers form an infinite set. There is a formula for odd numbers and that is **O = 2n + 1**, and the set of odd numbers is further obtained and this is done by substituting each of the integer numbers **I = {..., −3, −2, −1, 0, 1, 2, 3,...}** in turn for n.

Even Function

A function can be referred to as even, odd or neither of the two. There can be one simplest way to recognize a function if it is either even or odd and that is to look at its graph. The graph of an even function is considered symmetrical about the y-axis. There can be two examples such as $y = x^2 + 2$ and $y = \cos x$, and both functions are even due to the fact that their graphs are symmetrical about the y-axis.

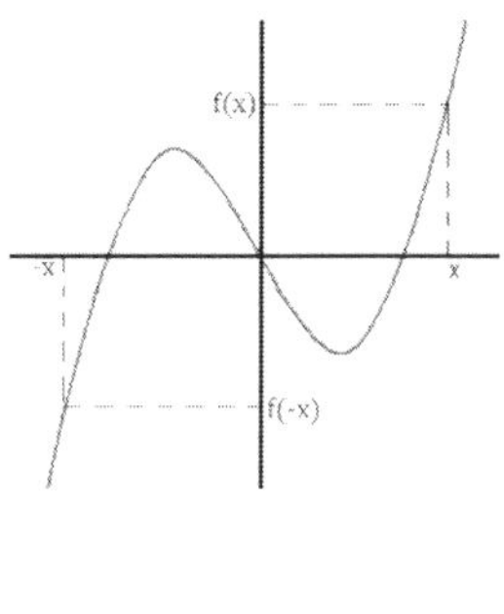

Figure 53. Even function.

Source: Image by Wikimedia commons.

Event

In probability, there are a number of terms which can be explained together. Let's take an example. William tosses two coins together a number of times, and then, records the results. This process is referred to as an experiment. One more example of an experiment can be rolling two dice together, or drawing a card from a pack of 52 playing cards. Considering an experiment there are a number of possible outcomes and there can be no way of predicting which outcome is next. The main aim of an experiment is to investigate the truth, or further, of a statement. William is performing his experiment in order to investigate the truth, or not, which when two coins are tossed together a head (H) and a tail (T) are more likely to turn up than two heads.

Expanding Brackets

The results formed from the process of expanding brackets are in an equivalent expression that contains no brackets. In simple words, it is to rewrite an expression for removing the brackets. The result formed will be an expression which can be referred to as the expansion of the brackets, and usually, it is longer in comparison to the original expression. There can be a practical approach in order to explain the expanding brackets, but when the process is understood, there can be use of a simple algorithm.

Exponent

However, when a number or an expression is multiplied by itself a lot of times it will be far easier to express the outcome in exponent form. For an example, $81 = 3 \times 3 \times 3 \times 3$ and can be written as 3^4. Thus, it can be said that 3^4 is the exponent form of $3 \times 3 \times 3 \times 3$ and the value is 81. Whereas, when 3^4 replaces 81, it is said that 3 is the base and 4 is the exponent. While, one more name for an exponent is index. On the other hand, the plural of index is indices.

The power can be referred to as the number of times the base is multiplied by itself. For example, 3^4 it can be said that 81 is the fourth power of 3, or 81 is 3 to the power 4. There are special names which are used for powers of 2 and 3. Considering an example, 4^2 is four squared or four to the power 2, and 5^3 is five cubed or five to the power 3. There can be one more way of expressing exponent form and that is as follows: 5 is raised to the power 4, that means 5^4. When numbers or algebraic terms in index form are either multiplied or divided, the application of specific rules can be done.

Exponential Curve

Sometimes, the exponential curve is referred to as a growth curve due to the fact that it models the way populations grow most of the time. This is shown as a curve whose equation is $y = a^x$. In this expression, a is any number that is greater than zero. Let's take an example, $y = 2^x$ is the equation of the exponential curve in the given figure. The curve has a horizontal asymptote which is the x – axis.

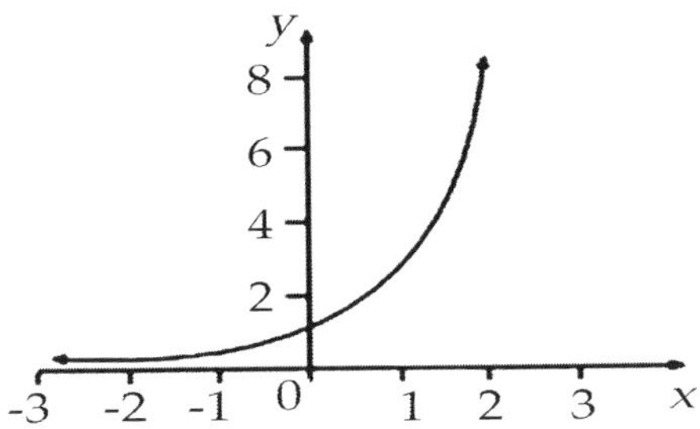

Figure 54. Exponential curve.

Source: Image by Wikipedia.

Exponential Decay

According to an exponential rule, the mass of a radioactive element decays over a period of time. The time taken in decaying process of half of its original mass is referred to as the half – life of the element. The mass m of an iodine element has a fast rate of decay. The original mass of the element is 80 grams and it decays to 40 grams in 8 days. This is half of its mass, so the half-life of the element is 8 days.

Exterior Angle of a Polygon

The exterior angle of a polygon can be defined as the angle between one side of a polygon and the extension of the next side. In the pentagon in the given figure, it is the shaded angle. A pentagon contains five exterior angles.

There is the geometry theorem about the exterior angles of a polygon which states:

"The sum of all the exterior angles of a polygon is 360°."

This theorem is said to be true for all polygons whether it contains 3 sides, 5 sides or even 100 sides.

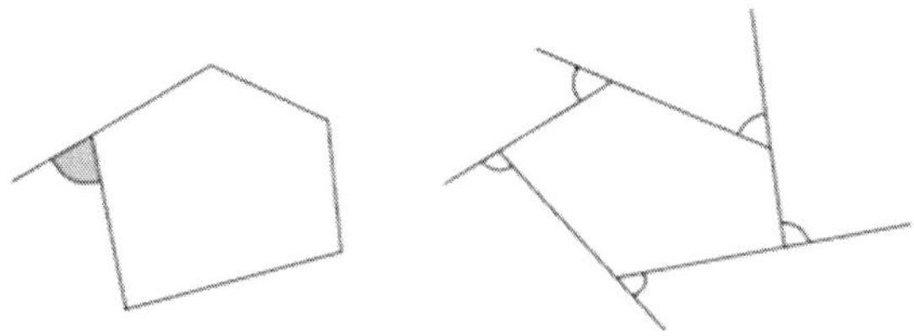

Figure 55. Exterior angle of a polygon.

Source: Image by Wikipedia.

Exterior Angle of a Triangle

For exterior angle of a triangle, the geometry theorem states that:

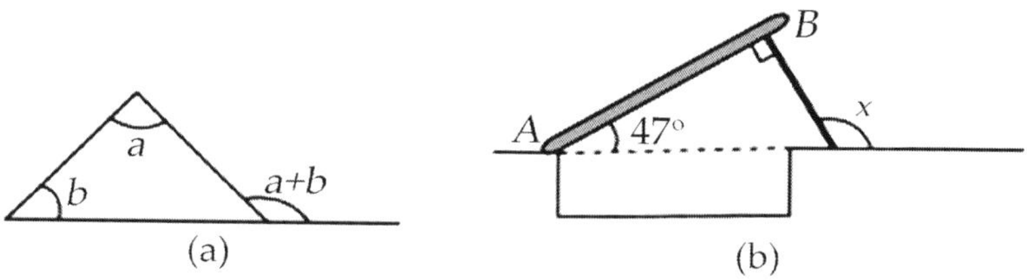

Figure 56. Exterior angle of a triangle.

Source: Image by Wikipedia.

"The exterior angle of a triangle is equal to the sum of the two interior opposite angles."

Extrapolation

The process of extrapolation is said to be a method of estimating the value of a function beyond the values which are already known. So as to extrapolate, the assumption can be done of the function which will continue in the same pattern that it has already followed. It is due to the fact that this assumption is made that extrapolation and interpolation are not absolutely reliable until unless one is certain that the pattern of the function will not change.

Expanded Form of Decimals

Expanded form is useful to split and present the higher digit number in its units, tens, hundreds, thousands form. An expanded form helps to better understand and rightly read the higher digit numbers. A number of the form 10030, is sometimes difficult to understand directly and can be represented in expanded form as 10030 = 10,000 + 30.

Unlike whole numbers, decimal numbers can also be written in the expanded form. For writing the decimals in expanded form, we multiply each of the decimal digits with increasing exponent values of (1/10). Let us try to understand this with the help of a simple example of a decimal number. The decimal number 0.437 can be written in expanded form as $4 \times (1/10) + 3 \times (1/10)^2 + 7 \times (1/10)^3 = 4 \times (1/10) + 3 \times (1/100) + 7 \times (1/1000) = 0.4 + 0.03 + 0.007$

Now with the possibility of expressing decimals in expanded form, we can write any number in expanded form. A fraction, a percentage value can be converted into a decimal and the same can be written in the expanded form. A fraction of 1/7 in the decimal form would be 0.1428, 0.1428 in the expanded form would be $0.1428 = 1 \times (1/10) + 4 \times (1/10)^2 + 2(1/10)^3 + 8(1/10)^4$. And a percentage of 25% would be $0.25 = 2 \times (1/10) + 5 \times (1/10)^2$

Experimental Probability

Experimental probability, also known as Empirical probability, is based on actual experiments and adequate recordings of the happening of events. To determine the occurrence of any event, a series of actual experiments are conducted. Experiments which do not have a fixed result are known as random experiments. The outcome of such experiments is uncertain. Random experiments are repeated multiple times to determine their likelihood. An experiment is repeated a fixed number of times and each repetition is known as a trial.

Experimental Probability Formula

The experimental probability of an event is based on the number of times the event has occurred during the experiment and the total number of times the experiment was conducted. Each possible outcome is uncertain and the set of all the possible outcomes is called the sample space. The formula to calculate experimental probability is: P(E) = Number of times an event occurs/Total number of times the experiment is conducted

Element

A member of, or an object in, a set.

Elliptic Geometry

a non-Euclidean geometry based (at its simplest) on a spherical plane, in which there are no parallel lines and the angles of a triangle sum to more than 180°.

Elliptic geometry is a geometry in which no parallel lines exist. In order to discuss the rigorous mathematics behind elliptic geometry, we must explore a consistent model for the geometry and discuss how the postulates posed by Euclid and amended by Hilbert must be adapted.

The most common and intuitive model of elliptic geometry is the surface of a sphere. In the spherical model a "point" is defined as a pair of antipodal points and a "line" is defined as a great circle of the sphere. This model may sound deceptively simple at first glance, but since this model has no parallel lines, it is not a model of neutral geometry.

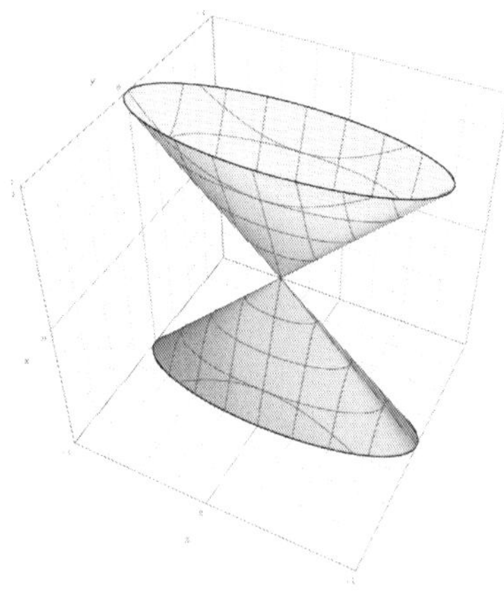

Figure 57. Elliptic geometry (cone quadric).

Source: Image by Wikimedia commons.

Empty (Null) Set

A set that has no members, and therefore has zero size, usually represented by {} or ø.

Euclidean Geometry

"Normal" geometry based on a flat plane, in which there are parallel lines and the angles of a triangle sum to 180°.

Euclidean geometry, the study of plane and solid figures on the basis of axioms and theorems employed by the Greek mathematician Euclid (c. 300 BCE). In its rough outline, Euclidean geometry is the plane and solid geometry commonly taught in secondary schools. Indeed, until the second half of the 19th century, when non-Euclidean geometries attracted the attention of mathematicians, *geometry* meant Euclidean geometry.

It is the most typical expression of general mathematical thinking. Rather than the memorization of simple algorithms to solve equations by rote, it demands true insight into the subject, clever ideas for applying theorems in special situations, an ability to generalize from known facts, and an insistence on the importance of proof. In Euclid's great work, the *Elements*, the only tools employed for geometrical constructions were the ruler and the compass—a restriction retained in elementary Euclidean geometry to this day.

In its rigorous deductive organization, the Elements remained the very model of scientific exposition until the end of the 19th century, when the German mathematician David Hilbert wrote his famous Foundations of Geometry (1899). The modern version of Euclidean geometry is the theory of Euclidean (coordinate) spaces of multiple dimensions, where distance is measured by a suitable generalization of the Pythagorean theorem.

Expected Value

The amount predicted to be gained, using the calculation for average expected payoff, which can be calculated as the integral of a random variable with respect to its probability measure (the expected value may not actually be the most probable value and may not even exist, e.g., 2.5 children).

F

Factor

A factor of a number is a positive integer that divides exactly into another positive integer without leaving a remainder. Every positive integer is a factor of 1, and every positive integer is a factor of itself. Because 8 divides perfectly into 24 without a leftover, it is a factor of 24. A factor of 24 is the integer 24, and 1 is also a factor of 24. 1, 2, 3, 4, 6, 8, 12, 24 is a complete list of the eight factors of 24.

A prime factor is a factor that has the same value as the prime number. The prime factors of 24 are {2, 3}. The integer 1 is not a prime number, but it is a factor of 24. A common factor of two numbers is a number that is a factor of both of them. Because 3 is a factor of both 6 and 9, it is a common factor of both 6 and 9.

Factor Tree

Factorizing numbers, or writing them as the products of their prime factors, is done using the factor tree.

Example: Write 840 as the product of primes.

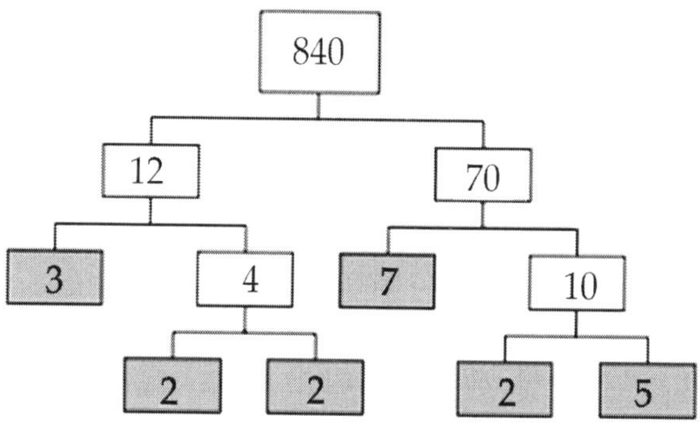

Factorial

In permutations and combinations, factorials are employed. The product of the first n counting numbers is the factorial of a counting number n. The factorial n sign is n! 0! = 1, and 1! = 1 by definition.

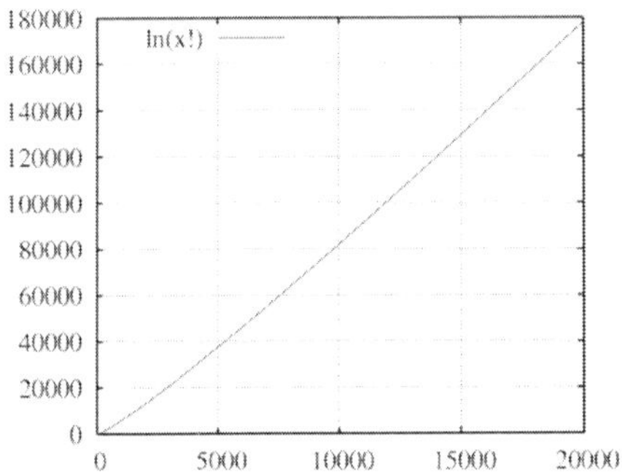

Figure 58. Factorial.

Source: Image by Wikimedia commons.

Factorize

When we factorize an equation, we use sets of brackets to write it as a product of its components. Factorizing is the inverse of expanding brackets, therefore it's important to understand expanding before moving on to factorizing. This entry covers three different types of factorization:

1. Common factors, type 1 and type 2.
2. Quadratic factors.
3. The difference between two squares.

Fibonacci Sequence

The Fibonacci number sequence is one of the most well-known in mathematics. Fibonacci (1180–1250), sometimes known as Leonardo of Pisa, was an Italian mathematician. The numbers 1, 1, 2, 3, 5, 8, 13, 21, 34, and so on make up the sequence. By combining the two prior terms, the following term is discovered. Because 55 = 21 + 34, the following term after 34 is 55. Fibonacci's number series began with the following difficulty that he was attempting to solve. "How many rabbit pairs will be produced in a year, starting with a single pair, if each pair bears a new pair every month, which becomes productive from the second month on?"

Formula

Formulae or formulas is the plural form of the word formula. A formula is a two- or more-quantity equation that shows how changes in one quantity impact the other (s). The area of a circle with a radius of R is calculated using the formula A = R2. The relationship between the two quantities A and R is represented by this formula. R affects A in the same way as it affects R. The area of a triangle with a base of length B and a perpendicular height of H is calculated using the formula A = 1/2BH. The relationship between the three quantities A, B, and H is represented by this formula. Changes in B and H cause an equal and opposite change in A.

Four-Color Problem

Let's say you want to color the different regions on a map, or any other figure with a lot of them. If you want to paint sections with a common edge differently, you'll never need more than four colors to color the entire map. Red ®, yellow (Y), blue (B), and green (G) have been used to color the map in the illustration (G). It's also necessary to color the surrounding area. Regions that meet at a point can be the same color, but those that meet on an edge must be different colors.

Four Rules

Adding, subtracting, multiplying, and dividing are the four rules. You'll need to know how to add, subtract, multiply, and divide decimals if you want to master the four decimal rules.

Fractions

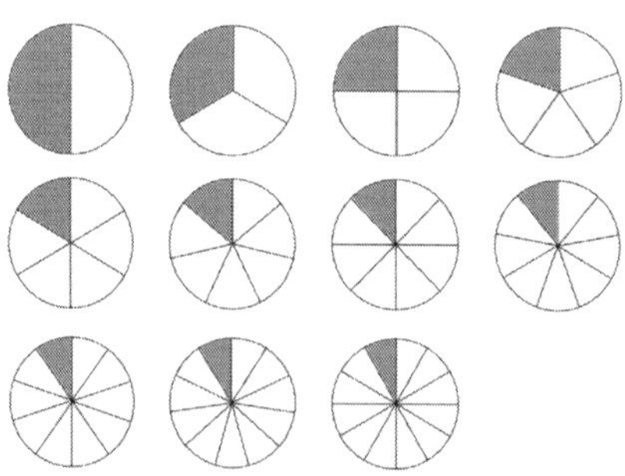

Figure 59. Fractions.

Source: Image by Flickr.

This entry will focus solely on numerical fractions. See the entries Algebraic Fractions and Canceling for more information on algebraic fractions. A fraction is a number formed by multiplying two integers by their ratio. If n and d are two numbers, then n d is a fraction if d is not equal to zero; otherwise, the fraction is undefined.

Frequency

The number of times an event occurs in an experiment is referred to as frequency in statistics. A frequency table, also known as a frequency distribution, is created by combining the frequencies of all the occurrences in an experiment and recording them in a table. These concepts are demonstrated in the example below, which uses a tally column to track the number of times an event occurs.

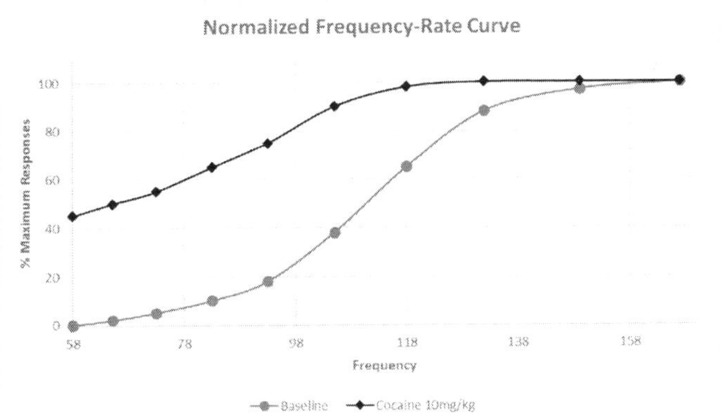

Figure 60. Frequency.

Source: Image by Wikimedia commons.

Frequency Polygon

This is a graph that resembles a histogram in appearance. In the following example, the significance of a frequency polygon will be demonstrated. Nathan gathered information on the weights of his 30 students in his class. He divided the weights into five-kilogram class intervals and recorded the results in a frequency chart, as displayed. The class interval 40– denotes a weight range of 40–45 kilograms, which includes 40 kilograms but excludes 45 kilograms. Nathan created a histogram of the findings using the frequency table. The histogram columns aren't part of the frequency polygon; therefore,

Furlong

A furlong is a unit of length in the US customary and imperial systems of measurement. The name is derived from the historical use of the furlong to refer to the average length of the furrow in 1 acre of a ploughed field, hence the name "furrow-long." 1 furlong is equal to:

1. mile
2. 660 feet
3. 220 yards
4. 40 rods
5. 201.168 meters (relative to international yard)
6 .201.1684026 meters (relative to US survey foot)

The definition of the furlong differs slightly depending on whether it is defined based on the international yard or the US survey foot. The international yard is defined as 0.9144 meters. The US does not formally define survey yards, but based on the survey foot, which is equal to 0.30480061 meters, the "survey yard" is 0.91440183 yards.

Practically, the difference in measurement in the furlong based on the difference between the international yard and survey foot is too small to result in significant consequences. Since survey measurements typically involve very large areas with variable conditions anyway, the best reported survey data is expected to have some degree of inaccuracy that would not really be distinguishable from any inaccuracy resulting from the minor difference in definitions of the furlong.

Since the furlong is no longer widely used, it can be useful to be able to convert from furlongs to more widely used units of measurement.

Function

The entry Correspondence contains a detailed description of the function. The relation tree depicts all subsets of relations, and there are two sorts of relations: functions and relationships.

G

Gears

A gear is a toothed wheel. To vary the speed and direction of a turning, two gear wheels are employed together. A bicycle has two gear wheels, one at the pedal and the other at the back wheel. The gear wheels are connected by a chain, which ensures that they turn in the same direction.

Figure 62. Gears.

Source: Image by Pxfuel.

Golden Ratio

The golden rectangle is another name for this shape. Assume that the point C divides a line segment AB into two unequal pieces. If the ratio of the complete segment AB to the bigger segment AC is equal to the ratio of the larger segment AC to the smaller segment CB, the point C splits the line segment in the golden ratio. This golden ratio expression can be written as AB/AC = AC/CB.

Gradient

Slope is another term for gradient. A straight line's gradient is stated as a fraction and indicates how steep the line is. Positive, negative, zero, or undefined gradients are possible. A line's gradient might be less than one or larger than one. The ideal way to describe the gradient of a straight line is to use squares on a grid, or x–y axes. The sign representing a straight line's gradient is m. For lines that slope in the direction '/', m is positive, while for lines that slope in the direction '\', m is negative. In other words, a positive gradient exists when a line increases from left to right, while a negative gradient exists when a line decreases from left to right.

Gradient-Intercept Form

This term refers to a technique for drawing straight-line graphs without the need of a table of values. Straight-line graph equations can be expressed as $y = mx + c$, which is known as the gradient-intercept version of the straight-line graph. A = is a graph that depicts a straight line. The straight-line graph's gradient is m, and its y intercept is c in the equation $y = mx + c$. To find the values of m and c, we must first ensure that the straight line's equation is expressed in the form $y = mx + c$ before drawing the graph. If the values of m and c are known, the graph can be drawn.

Gradient of A Curve

A curve's gradient varies along its length, therefore there is no single gradient, but rather a variable gradient at each point along the curve. We draw a tangent to the curve at that point and calculate the gradient of the tangent, which is defined to be the gradient of the curve at that point, to find the gradient at that position on a curve.

Gram

The gram (abbreviated as g) is a tiny mass unit equal to one thousandth of a kilogram (kg): 1000 g = 1 kg.

One key relationship is that 1 cubic centimeter of water (1 cm^3) weighs 1 gram when the temperature is 4 degrees Celsius (1 cm3 is the same as 1 milliliter). Thus, 1 cubic meter (1 m^3) of water weights 1000 kilograms, or 1 tons (abbreviation t; note that 1 t = 1000 kg = 2200 pounds, whereas 1 US ton = 2000 pounds).

Greater Than

The symbol for "greater than" is >. The number 5 is greater than the number −3, since 5 is to the right of −3 on the number line. This is written as $5 > -3$. The symbol for "less than" is < 5.

Gross

This is the number 144. Alternatively, one gross can be described as 12 dozen, where one dozen is 12.

Geometry Theorems

Geometry is a very organized and logical subject. The guiding light for solving Geometric problems is Definitions, Geometry Postulates, and Geometry Theorems. So before moving onto the geometry theorems list, let us discuss these to aid in geometry postulates and theorems list.

Angle Theorems

The relation between the angles that are formed by two lines is illustrated by the geometry theorems called "Angle theorems." Some of the important angle theorems involved in angles are as follows:

1. Alternate Exterior Angles Theorem

When two parallel lines are cut by a transversal then resulting alternate exterior angles are congruent.

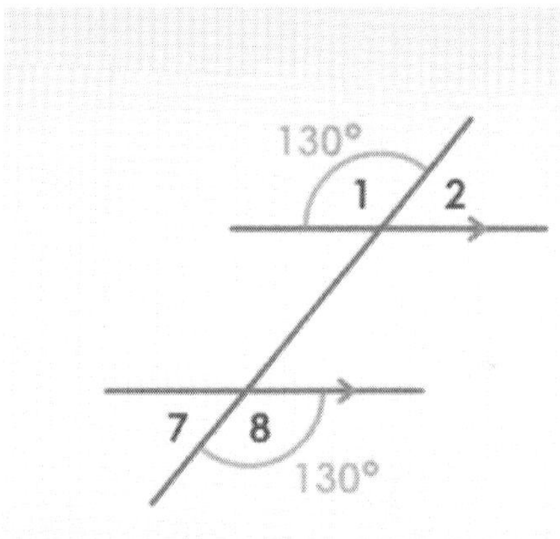

Figure 63. Alternate exterior angles theorem.

Source: Image by Flickr.

The alternate exterior angles have the same degree measures because the lines are parallel to each other.

2. Alternate Interior Angles Theorem

When two parallel lines are cut by a transversal then resulting alternate interior angles are congruent.

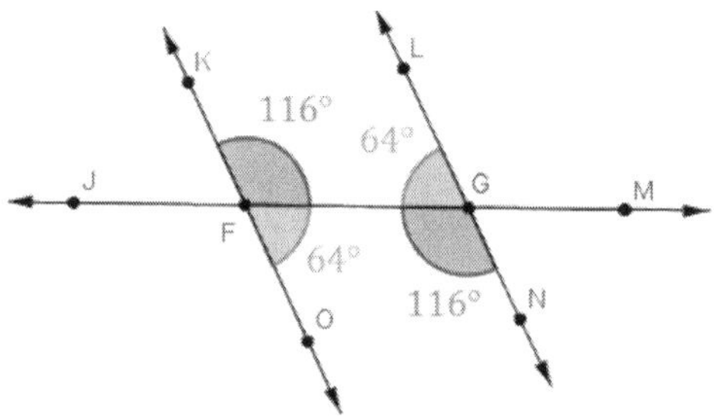

Figure 64. Alternate interior angles theorem.

Source: Image by Flickr.

The alternate interior angles have the same degree measures because the lines are parallel to each other.

One way to find the alternate interior angles is to draw a zig-zag line on the diagram.

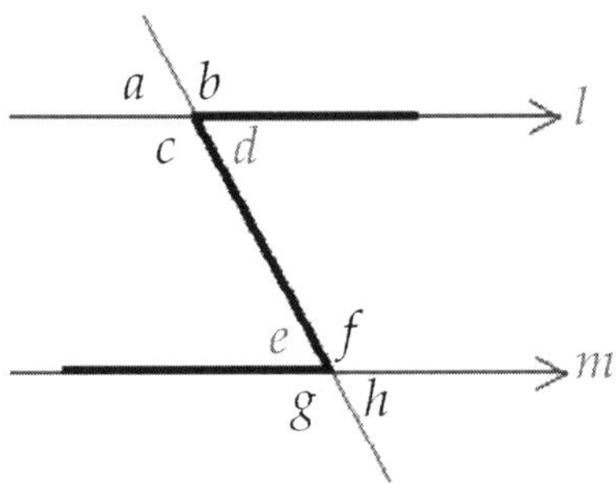

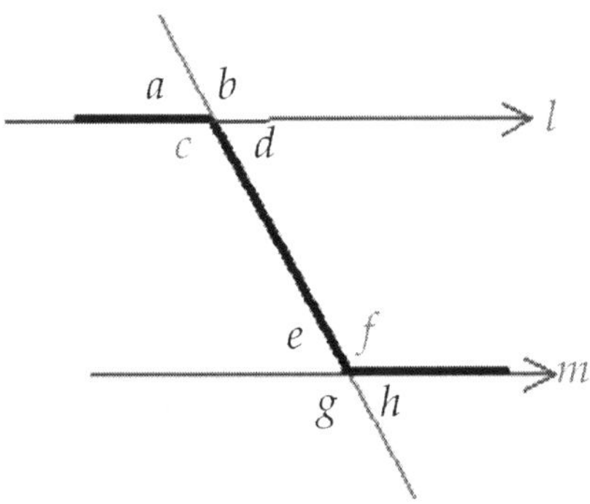

3. Congruent Complements Theorem

If two angles are complementary to the same angle or of congruent angles, then the two angles are congruent.

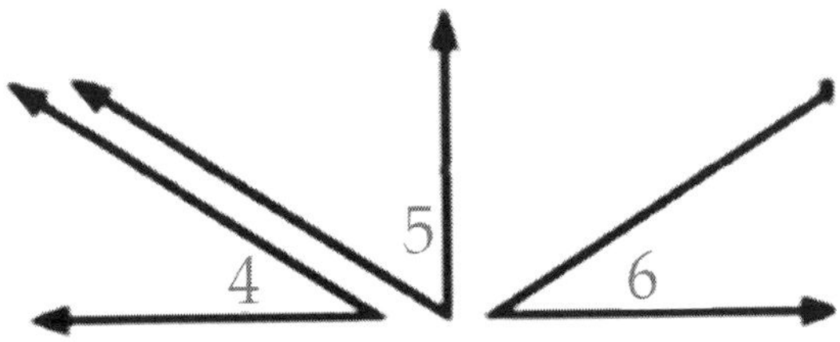

Figure 65. Congruent complements theorem.

Source: Image by Flickr.

4. Congruent Supplements Theorem

If two angles are supplements to the same angle or of congruent angles, then the two angles are congruent.

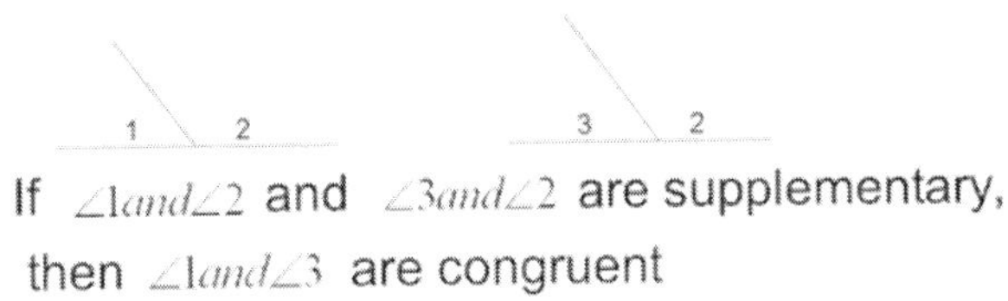

5. Right Angles Theorem

If two angles are both supplement and congruent then they are right angles.

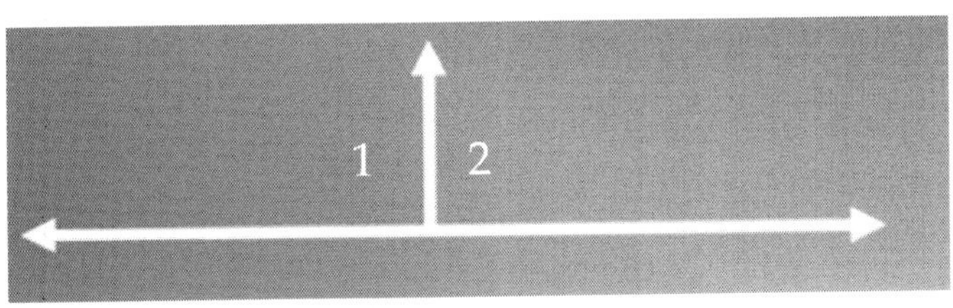

Figure 66. Right angles theorem.

Source: Image by Flickr

6. Same-Side Interior Angles Theorem

If two parallel lines are cut by a transversal, then the interior angles on the same side of the transversal are supplementary.

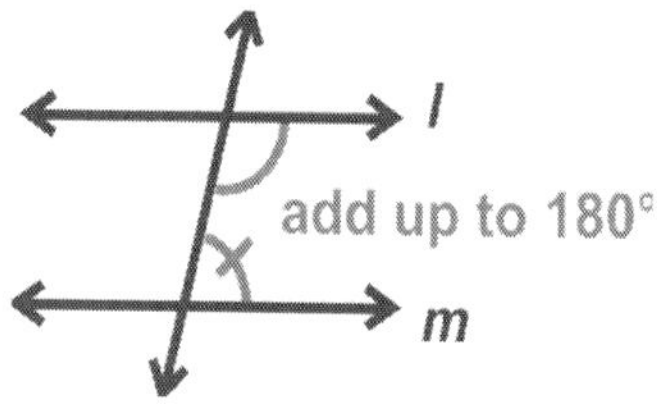

Figure 67. Same-side interior angles theorem.

Source: Image by Pixabay.

7. Vertical Angles Theorem

Angles that are opposite to each other and are formed by two intersecting lines are congruent.

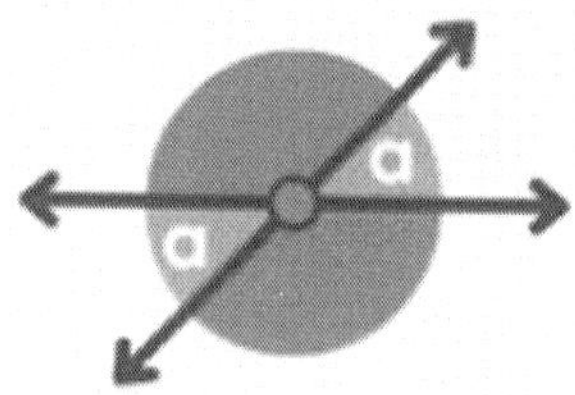

Figure 68. Vertical angles theorem.

Source: Image by Pixabay.

Now let us move onto geometry theorems which apply on triangles.

Triangle Theorems

We know that there are different types of triangles based on the length of the sides like a scalene triangle, isosceles triangle, equilateral triangle and we also have triangles based on the degree of the angles like the acute angle triangle, right-angled triangle, obtuse angle triangle.

Though there are many Geometry Theorems on Triangles but Let us see some basic geometry theorems.

Theorem 1

In any triangle, the sum of the three interior angles is 180°.

Theorem 2

If a side of the triangle is produced, the exterior angle so formed is equal to the sum of corresponding interior opposite angles.

Theorem 3

The base angles of an isosceles triangle are congruent.

Theorem 4

If a line divides any two sides of a triangle in the same ratio, then the line is parallel to the third side.

Theorem 5

If in two triangles, corresponding angles are equal, then their corresponding sides are in the same ratio and hence the two triangles are similar.

Theorem 6

If in two triangles, the sides of one triangle are proportional to other sides of the triangle, then their corresponding angles are equal and hence the two triangles are similar.

Circle Theorems

Circle theorems helps to prove the relation of different elements of the circle like tangents, angles, chord, radius, and sectors. Or we can say circles have a number of different angle properties, these are described as circle theorems.

Now let's study different geometry theorems of the circle.

Circle Theorems 1

Angles in the same segment and on the same chord are always equal.

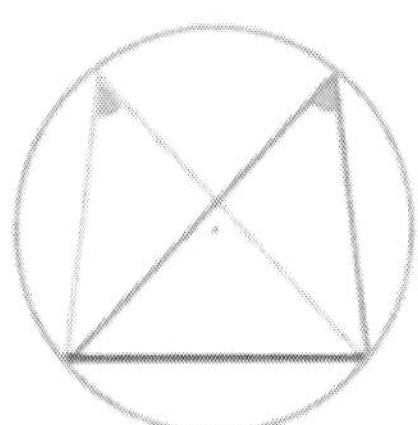

Circle Theorems 2

A line drawn from the center of a circle to the mid-point of a chord is perpendicular to the chord at 90°.

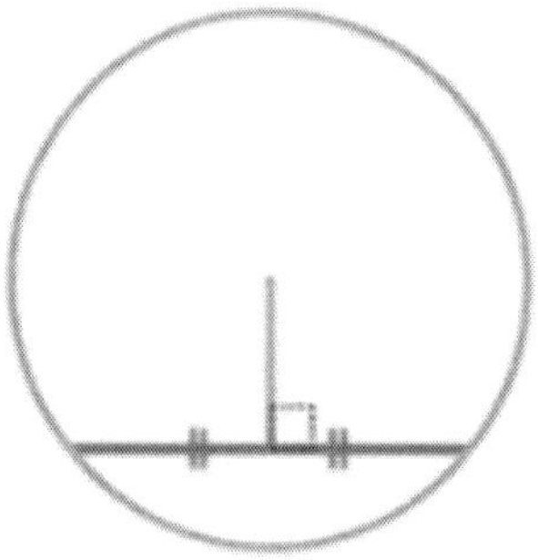

Circle Theorems 3

The angle at the center of a circle is twice the angle at the circumference.

Circle Theorems 4

The angle between the tangent and the side of the triangle is equal to the interior opposite angle.

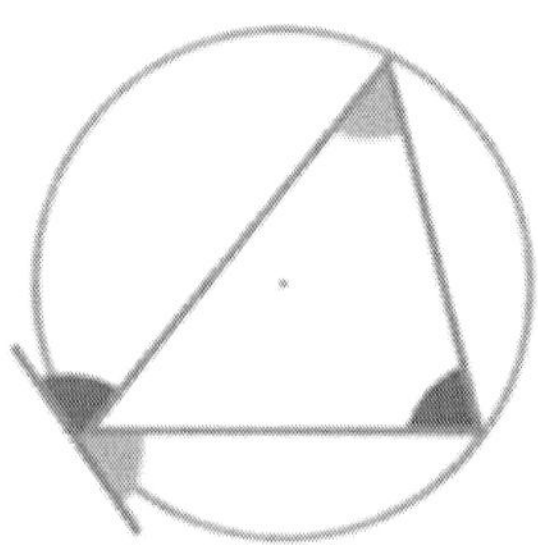

Circle Theorems 5

The angle in a semi-circle is always 90°.

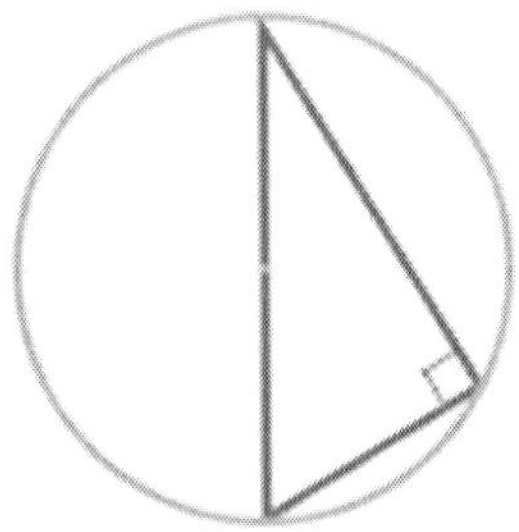

Circle Theorems 6

Tangents from a common point (A) to a circle are always equal in length. AB=BC

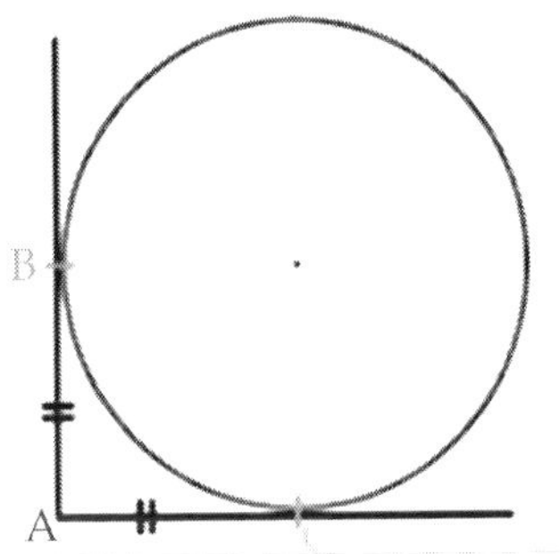

Circle Theorems 7

The angle between the tangent and the radius is always 90°

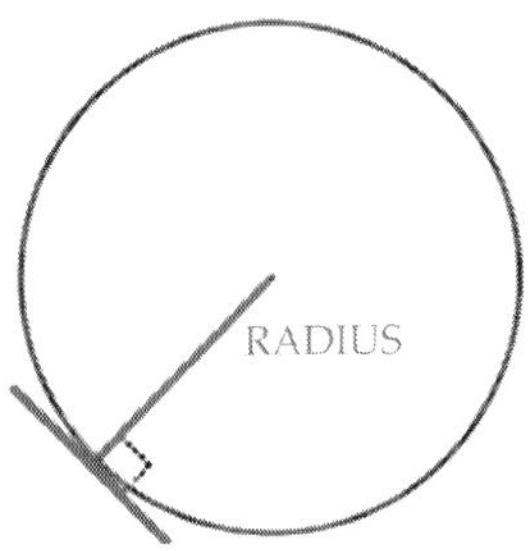

Circle Theorems 8

In a cyclic quadrilateral, all vertices lie on the circumference of the circle. Opposites angles add up to 180°.

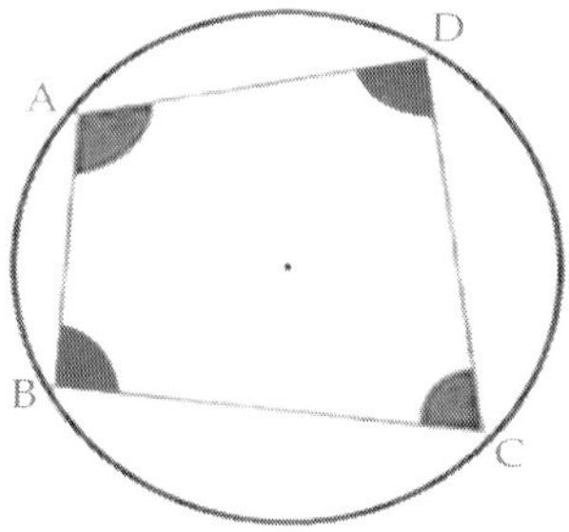

Proceed to the discussion on geometry theorems dealing with parallelograms or parallelogram theorems.

Parallelogram Theorems

A parallelogram is a quadrilateral with both pairs of opposite sides parallel.

Let's now understand some of the parallelogram theorems.

Parallelogram Theorems 1

If both pairs of opposite sides of a quadrilateral are congruent, then the quadrilateral is a parallelogram.

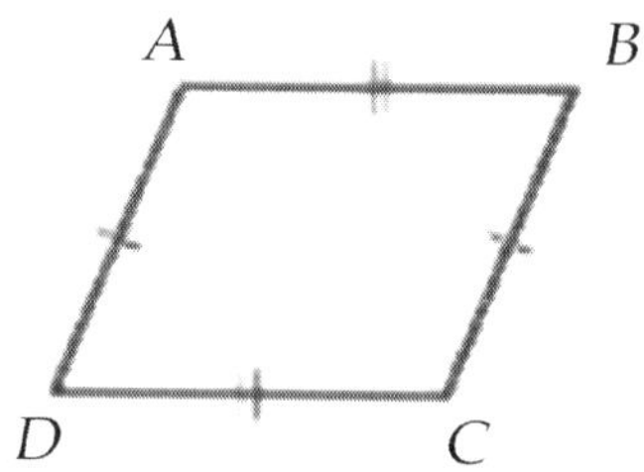

Parallelogram Theorems 2

If both pairs of opposite angles of a quadrilateral are congruent, then the quadrilateral is a parallelogram.

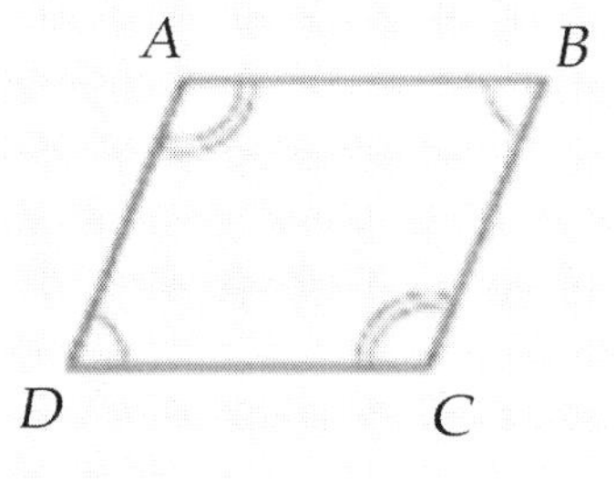

Parallelogram Theorems 3

If the diagonals of a quadrilateral bisect each other, then the quadrilateral is a parallelogram.

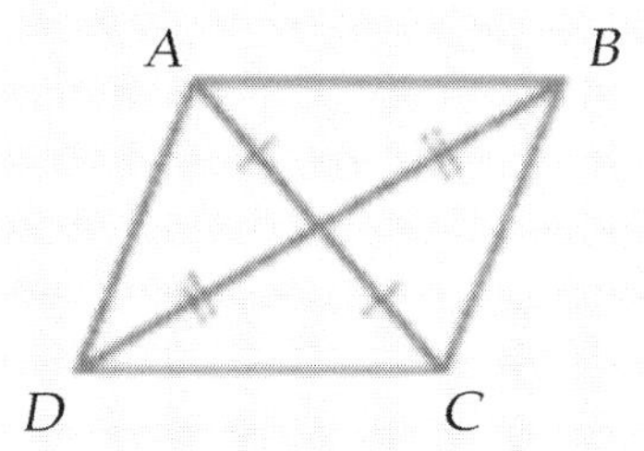

Parallelogram Theorems 4

If one pair of opposite sides of a quadrilateral is both parallel and congruent, then the quadrilateral is a parallelogram.

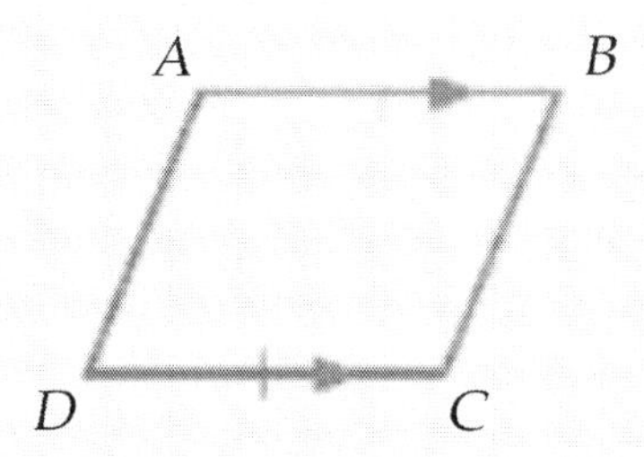

Growth Curve

A growth curve is a graphical representation of how a particular quantity increases over time. Growth curves are used in statistics to determine the type of growth pattern of the quantity—be it linear, exponential, or cubic.

Once the type of growth is determined, a business can create a mathematical model to predict future sales. An example of a growth curve is a country's population over time.

The shape of the growth curve can make a big difference when businesses determine whether to launch a new product or enter a new market. Slow growth markets are less likely to be appealing because there is less room for profit, while exponential growth could mean that the market could see a lot of competitors enter the market.

Growth curves began in the physical sciences such as biology; today, they're a common component to social sciences as well. Advancements in digital technologies and business models now require analysts to account for growth patterns unique to the modern economy.

For example, the winner-take-all phenomenon is a fairly recent development brought on by the likes of Amazon, Google, and Apple. Researchers are scrambling to make sense of growth curves unique to their business models and platforms.

Future shifts in demographics (aging), the nature of work, and artificial intelligence will further strain conventional ways of analyzing growth curves or trends.

The expression growth curve might be considered more reserved for scientific approaches to what the layman may describe as growth trends. Definition aside, growth curves play an essential role in determining the future success of products, markets, and societies, both at the micro and macro levels.

Game Theory

A branch of mathematics that attempts to mathematically capture behavior in strategic situations, in which an individual's success in making choices depends on the choices of others, with applications in the areas of economics, politics, biology, engineering, etc.

Game theory, branch of applied mathematics that provides tools for analyzing situations in which parties, called players, make decisions that are interdependent. This interdependence causes each player to consider the other player's possible decisions, or strategies, in formulating strategy. A solution to a game describes the optimal decisions of the players, who may have similar, opposed, or mixed interests, and the outcomes that may result from these decisions.

Although game theory can be and has been used to analyze parlor games, its applications are much broader. In fact, game theory was originally developed by the Hungarian-born American mathematician John von Neumann and his Princeton University colleague Oskar Morgenstern, a German-born American economist, to solve problems in economics.

In their book *The Theory of Games and Economic Behavior* (1944), von Neumann and Morgenstern asserted that the mathematics developed for the physical sciences, which describes the workings of a disinterested nature, was a poor model for economics. They observed that economics is much like a game, wherein players anticipate each other's moves, and therefore requires a new kind of mathematics, which they called game theory. (The name may be somewhat of a misnomer—game theory generally does not share the fun or frivolity associated with games.)

Game theory has been applied to a wide variety of situations in which the choices of players interact to affect the outcome. In stressing the strategic aspects of decision making, or aspects controlled by the players rather than by pure chance, the theory both supplements and goes beyond the classical theory of probability.

It has been used, for example, to determine what political coalitions or business conglomerates are likely to form, the optimal price at which to sell products or services in the face of competition, the power of a voter or a bloc of voters, whom to select for a jury, the best site for a manufacturing plant, and the behavior of certain animals and plants in their struggle for survival. It has even been used to challenge the legality of certain voting systems.

Gaussian Curvature

An intrinsic measure of the curvature of a point on a surface, dependent only on how distances are measured on the surface and not on the way it is embedded in space.

In differential geometry, the Gaussian curvature or Gauss curvature K of a surface at a point is the product of the principal curvatures, κ_1 and κ_2, at the given point:

$$K = K_1 K_2$$

The Gaussian radius of curvature is the reciprocal of K. For example, a sphere of radius r has Gaussian curvature $1/r^2$ everywhere, and a flat plane and a cylinder have Gaussian curvature zero everywhere. The Gaussian curvature can also be negative, as in the case of a hyperboloid or the inside of a torus. Gaussian curvature is an *intrinsic* measure of curvature, depending only on distances that are measured on the surface, not on the way it is isometrically embedded in Euclidean space. This is the content of the Theorema egregium.

Gaussian curvature is named after Carl Friedrich Gauss, who published the Theorema egregium in 1827.

Geometry

The part of mathematics concerned with the size, shape and relative position of figures, or the study of lines, angles, shapes and their properties

Graph Theory

A branch of mathematics focusing on the properties of a variety of graphs (meaning visual representations of data and their relationships, as opposed to graphs of functions on a Cartesian plane).

Graph Theory, in discrete mathematics, is the study of the graph. A graph is determined as a mathematical structure that represents a particular function by connecting a set of points. It is used to create a pairwise relationship between objects.

The graph is made up of vertices (nodes) that are connected by the edges (lines). The applications of the linear graph are used not only in Math's but also in other fields such as Computer Science, Physics and Chemistry, Linguistics, Biology, etc. In real-life also the best example of graph structure is GPS, where you can track the path or know the direction of the road.

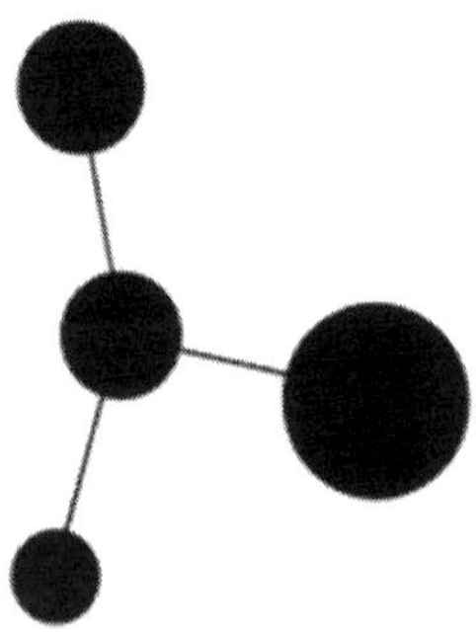

Figure 69. Graph theory.

Source: Image by Noun project.

Group

A mathematical structure consisting of a set together with an operation that combines any two of its elements to form a third element, e.g., the set of integers and the addition operation form a group

Group Theory

The mathematical field that studies the algebraic structures and properties of groups and the mappings between them.

Group theory, in modern algebra, the study of groups, which are systems consisting of a set of elements and a binary operation that can be applied to two elements of the set, which together satisfy certain axioms. These require that the group be closed under the operation (the combination of any two elements produces another element of the group), that it obey the associative law, that it contain an identity element (which, combined with any other element, leaves the latter unchanged), and that each element have an inverse (which combines with an element to produce the identity element).

If the group also satisfies the commutative law, it is called a commutative, or abelian, group. The set of integers under addition, where the identity element is 0 and the inverse is the negative of a positive number or vice versa, is an abelian group.

Groups are vital to modern algebra; their basic structure can be found in many mathematical phenomena. Groups can be found in geometry, representing phenomena such as symmetry and certain types of transformations. Group theory has applications in physics, chemistry, and computer science, and even puzzles like Rubik's Cube can be represented using group theory.

H

Hectare

This is a unit for measuring land areas such as farms, playing fields, school grounds, parks, etc. The abbreviation for hectare is ha. One hectare is equal in size to 10,000 square meters, which is a square measuring 100 meters by 100 meters. Roughly, two average soccer pitches together have the same area as 1 hectare.

Height

This is the vertical distance between an object's base and its top. Alternatively, height can be defined as an object's altitude. We must first determine the heights of various shapes before we can determine their areas and volumes. The length S is the slant height, or sloping height. Assume the triangle is twisted, and we need to know the height and base of the triangle to calculate its area. The height is still the distance between the top and the bottom measured perpendicularly.

Helix

A helix is a three-dimensional curve created by wrapping a right-angled triangle around a cylinder without overlapping it.

Hemisphere

A hemisphere is half of a sphere that is created when a sphere is sliced in half evenly. The cathedral's dome is a hemisphere. A hemisphere's volume is half that of a whole sphere, however determining the surface area of a hemisphere requires extra caution.

Hendecagon

This is an endecagon, which is a polygon with 11 sides.

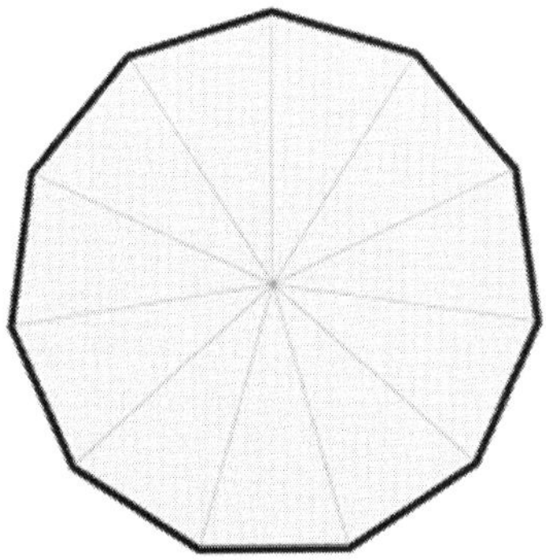

Figure 70. Hendecagon.

Source: Image by Flickr.

Hendecahedron

This is an 11-faced solid.

Heptagon

This is a seven-sided polygon and is also called a septagon.

Heron's Formula

Heron of Alexandria (also known as Hero) was a Greek mathematician who developed a formula for calculating the area of any triangle given the length of each side. This formula is written in the same notation that is used in trigonometry, and it is briefly described here. The lengths of the sides of the triangle opposite the vertices are called a, b, and c, respectively, and the vertices of a triangle are designated A, B, and C. The triangle's perimeter is a + b + c. s = (a + b + c)/2 gives half of the triangle's perimeter (semi perimeter).

The area of any triangle is calculated using Heron's formula.

1. A = s (s – a)
2. (s – b)
3. (s – c)

Hexagon

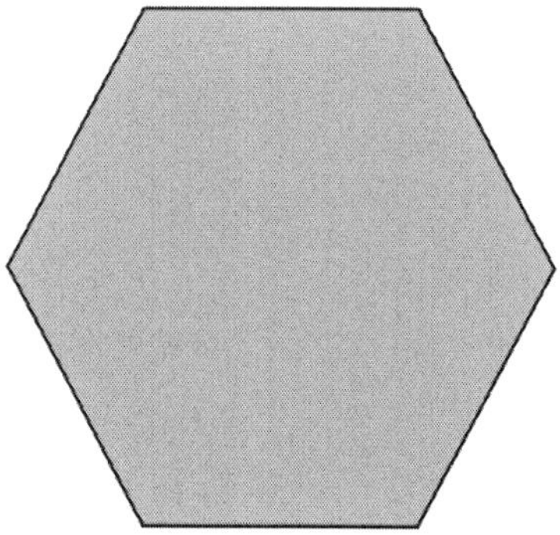

Figure 71. Hexagon.

Source: Image by Wikimedia commons.

A hexagon is a six-sided polygon. The regular hexagon has six equal-length sides and six equal-sized angles, totaling 120. It has six symmetry axes and a rotational symmetry order of six. Tessellation refers to the formation of a tiling pattern or mosaic from a regular hexagon. Bees employ this tiling pattern to construct their honeycomb. Six equilateral triangles make create a regular hexagon.

Hexagram

When all of the sides of a regular hexagon, , are extended, a six-pointed star known as a hexagram results.

Figure 72. Hexagram.

Source: Image by Flickr.

Hexahedron

A cube is sometimes known as a hexahedron, which is a regular polyhedron having six faces.

Hexomino

A hexomino is formed when six squares are linked edge to edge. Each pair of squares must be joined along the entire edge. There are 35 possible ways to put six squares together, and each one is referred to as a hexomino. Some of the hexominoes are cube nets. Keep in mind that the kinds in figure an are all considered to be the same hexomino.

Hire Purchase

When we want to buy something from a store but don't want to pay for it all at once, we use hire purchase (or "purchasing on time") to pay for it in installments. You will be able to take the item with you if you pay a portion of the money up front, known as the deposit. For example, let's say you want to spend \$700 on a CD player. You will be required to make an immediate payment of, say, 10% of the total price, or \$70. This is referred to as the CD player deposit.

Then, over the course of, say, two years, you'll be required to reimburse the remaining \$630 in monthly installments. However, because he has to wait two years for all of his money, the shop will add additional interest to the \$630. It's possible that the interest will be 15% of the outstanding balance. Now, 15% of \$630 becomes \$94.50, which is the interest charged by the shop.

So, the total amount you owe is \$630 + \$94.50 = \$724.50, which you must pay in installments. If the installments are made monthly, there will be 24 payments made over the course of two years. Each payment will be \$724.50 divided by 24 for a total of \$30.1875, which can be rounded up to \$30.20.

Highest Common Factor

The Highest Common Factor (HCF) of two numbers is the highest possible number which divides both the numbers exactly. HCF of a and b is denoted by HCF (a, b). Let d be the HCF (a, b), then you can't find the common factor of a and b greater than the number d. The highest common factor (HCF) is also called the greatest common divisor (GCD).

The HCF (Highest Common Factor) of two or more numbers is the highest number among all the common factors of the given numbers or in simple words, the HCF (Highest Common Factor) of two natural numbers x and y is the largest possible number that divides both x and y. Let's understand this definition using two numbers, 18 and 27. The common factors of 18 and 27 are 1, 3, and 9. Among these numbers, 9 is the highest (largest) number. So, the HCF of 18 and 27 is 9. This is written as: HCF (18,27) = 9

Histogram

A histogram is a frequency distribution statistical graph. It looks like a bar graph, but there are some differences. Because a histogram is a graph of continuous data, there are no spaces between the columns. The area of the column represents the frequency in theory; therefore, the columns do not have to be equal widths, but in fact, the columns of the histogram are normally equal width, and the frequency is then represented by the column height. How to draw a histogram is demonstrated in the example below.

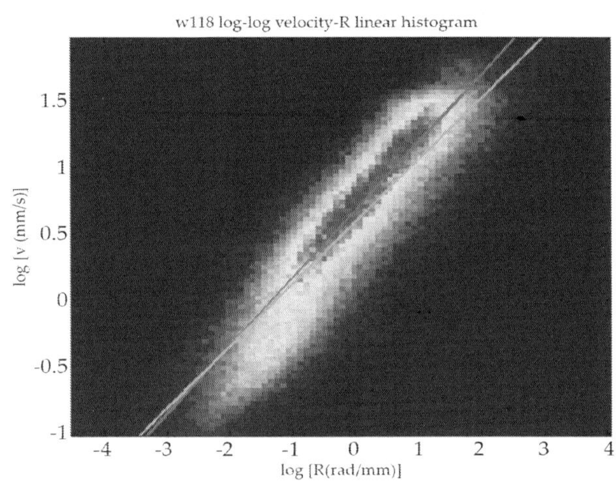

Figure 73. Histogram.

Source: Image by Wikipedia.

Horizontal

The horizontal, often known as level, is perpendicular to the vertical. A plumb line is a tool that can be used to define vertical. A plumb bob is a length of thread with one end attached to a small weight. The string's other end is attached to something, such as a beam, and if the small weight is allowed to hang freely without swinging, the string will be vertical. The beam will be horizontal if it is at right angles to the string.

Hyperbola

This is a curve made up of two independent branches, sometimes referred to as arms. A hyperbola's equation is of the kind $y = a/x$, or it can be expressed as $xy = a$, where an is a positive or negative number. Asymptotes are found on a hyperbola. They aren't actually part of the graph, but they're generally drawn to go with it.

Hypotenuse

The hypotenuse is the name of the side opposite the right angle of a right-angled triangle. The hypotenuse is always the longest side in a right-angled triangle.

I

Icosahedron

This is a polyhedron with 20 faces, as indicated by the prefix icosa. The icosahedron contains 20 congruent faces, each of which is an equilateral triangle. One of the Platonic solids is a regular icosahedron. A regular icosahedron with its net, which is made up of 20 equilateral triangles.

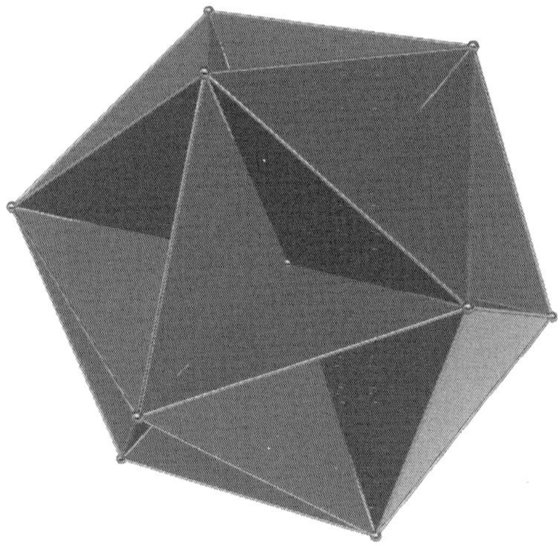

Figure 74. Icosahedron.

Source: Image by Wikimedia commons.

Image

An image is the result of applying a transformation on a shape, often known as an object.

Imperial System of Units

The yard is used to measure length, and the pound is used to measure weight. Inch, foot, yard, and mile are the most common length units. The following are the relationships between the length units:

12 inches = 1 foot

3 feet = 1 yard

22 yards = 1 chain

10 chains = 1 furlong

22 furlongs = 1 mile

1760 yards = 1 mile

The main units of weight are ounce (oz), pound (lb), hundredweight (cwt), and ton:

16 oz = 1 LB

112 LB = 1 cwt

20 cwt = 1 ton

2240 LB = 1 ton

The units for capacity of liquids are pint, quart, and gallon:

2 pints = 1 quart

4 quarts = 1 gallon

8 pints = 1 gallon

In terms of area, one acre has 4840 square yards.

Incenter

The incenter of a triangle is the center of the inscribed circle, also known as the incircle of a triangle. The point in the diagram is the triangle's incenter. A triangle's inscribed circle is a circle that is inside the triangle and touches each of its three sides. The inradius is the radius of the incircle. There is only one incenter in each triangle since there is only one incircle. The triangle's three sides are tangents to its incircle.

Inclined Plane

This plane is neither horizontal nor vertical. Because the home is sloping, the roof is depicted as an inclined plane. The house's wall is a vertical plane, while the house's floor is a horizontal plane.

Figure 75. Inclined plane.

Source: Image by Noun project.

Inconsistent Equations

If two equations can't be true at the same time, they're incompatible. We can alternatively state that the two equations are mutually exclusive. For instance, consider the following two equations.

$2x + 3y = 5$ and $2x + 3y = 6$

Are incompatible equations because they can't both be true at the same time: At the same time, $2x + 3y$ cannot be equal to two separate numbers. Let's pretend Nathan and Jacob went to a café for lunch. Nathan's bill was $5 for two cakes and three buns. Jacob ordered the same stuff as the others, two cakes and three buns, for $6. If each person had the same type of buns and cakes, the results would be inconsistent.

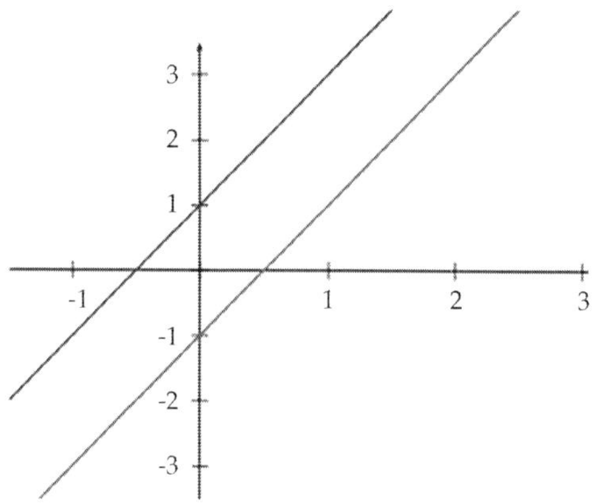

Figure 76. Inconsistent equations.

Source: Image by Flickr.

Increasing Function

A function is called increasing on an interval if given any two numbers, and in such that , we have.

Similarly, is called decreasing on an interval if given any two numbers, and in such that , we have.

Independent Events

Independent events are those events whose occurrence is not dependent on any other event. For example, if we flip a coin in the air and get the outcome as Head, then again if we flip the coin but this time we get the outcome as

Tail. In both cases, the occurrence of both events is independent of each other. It is one of the types of events in probability.

In Probability, the set of outcomes of an experiment is called events. There are different types of events such as independent events, dependent events, mutually exclusive events, and so on.

If the probability of occurrence of an event A is not affected by the occurrence of another event B, then A and B are said to be independent events.

Consider an example of rolling a die. If A is the event 'the number appearing is odd' and B be the event 'the number appearing is a multiple of 3,' then

$P(A)= 3/6 = 1/2$ and $P(B) = 2/6 = 1/3$

Also, A and B is the event 'the number appearing is odd and a multiple of 3' so that

$P(A \cap B) = 1/6$

$P(A | B) = P(A \cap B)/ P(B)$

$= (1/6)/(1/3) = 12$

$P(A) = P(A | B) = 1/2$, which implies that the occurrence of event B has not affected the probability of occurrence of the event A.

If A and B are independent events, then $P(A | B) = P(A)$

Using Multiplication rule of probability, $P(A \cap B) = P(B).P(A | B)$

$P(A \cap B) = P(B).P(A)$

Indices

Index (indices) in Math's is the power or exponent which is raised to a number or a variable. For example, in number 2^4, 4 is the index of 2. The plural form of index is indices. In algebra, we come across constants and variables. The constant is a value which cannot be changed. Whereas a variable quantity can be assigned any number or we can say its value can be changed. In algebra, we deal with indices in terms of numbers. A number or a variable may have an index. Index of a variable (or a constant) is a value that is raised to the power of the variable. The indices are also known as **powers or exponents**. It shows the number of times a given number has to be multiplied. It is represented in the form:

$a^m = a \times a \times a \times ......\times a$ (m times)

Here, a is the base and m is the index.

The index says that a particular number (or base) is to be multiplied by itself, the number of times equal to the index raised to it. It is a compressed method of writing big numbers and calculations.

Example: $2^3 = 2 \times 2 \times 2 = 8$

In the example, 2 is the base and 3 is the index.

Laws of Indices

There are some fundamental rules or laws of indices which are necessary to understand before we start dealing with indices. These laws are used while performing algebraic operations on indices and while solving the algebraic expressions, including it.

Rule 1: If a constant or variable has index as '0,' then the result will be equal to one, regardless of any base value.

$a^0 = 1$

Example: $5^0 = 1$, $12^0 = 1$, $y^0 = 1$

Rule 2: If the index is a negative value, then it can be shown as the reciprocal of the positive index raised to the same variable.

$a^{-p} = 1/a^p$

Example: $5^{-1} = {}^1/_5$, $8^{-3} = 1/8^3$

Rule 3: To multiply two variables with the same base, we need to add its powers and raise them to that base.

$a^p.a^q = a^{p+q}$

Example: $5^2.5^3 = 5^{2+3} = 5^5$

Rule 4: To divide two variables with the same base, we need to subtract the power of denominator from the power of numerator and raise it to that base.

$a^p/a^q = a^{p-q}$

Example: $10^4/10^2 = 10^{4-2} = 10^2$

Rule 5: When a variable with some index is again raised with different index, then both the indices are multiplied together raised to the power of the same base.

$(a^p)^q = a^{pq}$

Example: $(8^2)^3 = 8^{2.3} = 8^6$

Rule 6: When two variables with different bases, but same indices are multiplied together, we have to multiply its base and raise the same index to multiplied variables.

$\mathbf{a^p.b^p = (ab)^p}$

Example: $3^2.5^2 = (3 \times 5)^2 = 15^2$

Rule 7: When two variables with different bases, but same indices are divided, we are required to divide the bases and raise the same index to it.

$\mathbf{a^p/b^p = (a/b)^p}$

Example: $3^2/5^2 = ({}^3/_5)^2$

Rule 8: An index in the form of a fraction can be represented as the radical form.

$\mathbf{a^{p/q} = \sqrt[q]{a^p}}$

Example: $6^{1/2} = \sqrt{6}$

Inverse Proportion

Two quantities are said to be inversely proportional when the value of one quantity increases with respect to a decrease in another or vice-versa. This means that these two quantities behave opposite in nature. For example, the time taken to complete a task decreases with the increase in the number of workers finishing it and would increase with the decrease in the number of workers. Here, time and number of workers are inversely proportional to each other.

The other terms that can be used here for this type of proportion are inverse proportion or varying inversely or inverse variation or reciprocal proportion. Two variables say x and y, which are in inverse proportion are represented as $x \propto 1/y$ or $x \propto y^{-1}$. Direct proportion and inverse proportion are opposite relations in comparison to one another.

What is Inversely Proportional?

In Mathematics and Physics, we learn about quantities that depend upon one another, and such quantities are termed as proportional to one another. In other words, two variables or quantities are proportional to each other, if one is varied, then the other also changes by a fixed amount. This property of variables is termed proportionality and the symbol used to represent the proportionality is "$\propto$." There are two types of proportionality of variables.

They are:

- Directly Proportional
- Inversely Proportional

When two quantities are related to each other inversely, i.e., when an increase in one quantity brings a decrease in the other and vice versa then they are said to be inversely proportional. In inverse proportion, if one variable decreases, the other increases in the same proportion. It is opposite to direct proportion. Or, two quantities are said to be inversely proportional when one quantity is in direct proportion to the reciprocal of other. For example, the relation between speed and time. Speed and travel time are inversely proportional because the faster we travel, the lesser is the time taken, i.e., greater the speed, the shorter the time.

- As speed increases, travel time decreases
- And as speed decreases, travel time increases

Figure 77. Inversely proportional.

Source: Image by Flickr.

Inverse Trigonometric Ratios

Inverse trigonometric ratios are the trigonometric ratios that are used to find the value of the unknown angle with a given measure of the ratio of sides of the right-angled triangle. As we have used angles to find the trigonometric ratios of the sides of the triangle, similarly we can use the trigonometric ratios to find the angle. For example, $\sin(\theta) = (\text{Opposite})/(\text{Hypotenuse})$, hence we can get angle as $\sin^{-1}(\text{Opposite})/(\text{Hypotenuse}) = \theta$.

Using inverse trigonometric ratios, we have found out the inverse of the trigonometric function($\sin^{-1}x$) which is not the same as 1/sin x. The inverse is indicative inverse and is not an exponent. Inverse trigonometric ratios are the inverse of the trigonometric functions operating on the ratio of the sides of the triangle to find out the measure of the angles of the right-angled triangle. The inverse of a function is denoted by the superscript "–1" of the given trigonometric function. For example, the inverse of the cosine function will be $\cos^{-1}$. The inverse of the trigonometric function is also written as an "arc"-trigonometric function, for example, arc sin will be the inverse of the sine function. Inverse trigonometric ratios are used when we have the measure of the sides of the right-angled triangle and want to know the measure of the angles of the triangle.

Indirect Transformation

Two types of congruences are described under the heading Congruent Figures: Direct conversions that are directly congruent Indirect changes that are diametrically opposed.

The object and the image are directly congruent in a direct transformation, while the object and the image are oppositely congruent in an indirect transformation. Reflection is an indirect transformation, while rotation, translation, and enlargement are direct transformations. Another method of determining if a transformation is direct or indirect is to examine the object and picture as follows: The triangle ABC is reflected in the mirror line m in the figure, and its image is triangle ABC. You can move counterclockwise by moving your finger around the triangle ABC in the alphabetical sequence of the letters. Moving your finger around the image triangle ABC in alphabetical order, on the other hand, will take you in a clockwise direction. All indirect transformations have this property. The alphabetical direction described here is not reversed for any direct transformations.

Inequality

The graphs of the following inequalities are examined in this entry:

- Greater than (>)
- a number that is greater than or equal to ()
- Less than (<)
- a value that is less than or equal to ()

Inequations

An inequation is identical to an equation, except that instead of an equal sign, an inequality sign connects two expressions. The equation $2x + 5 = 3$ can be written as an inequation by replacing = by ≤ to give $2x + 5 \leq 3$. Inequations can be solved using the same techniques as are used to solve equations (refer to the entry Balancing an Equation). The contrast between balancing equations and balancing inequations is significant. When multiplying (or dividing) both sides of an inequation by a negative value, the inequality sign must be turned around before proceeding.

Infinite

It is not finite if anything is limitless. The term "infinity" refers to the size of a quantity (which can be a set, sequence, or group) that has no limit. It has no limits in terms of size; it is limitless. Because there is no last number in the sequence, the set of positive even numbers 2, 4, 6, 8,... is infinite. A line has an unlimited length in geometry; however, a line segment has a finite length.

Infinite Decimals

There is no limit to the number of decimal places in these decimals. Infinite decimals can be divided into two categories:

- Recurring decimals, such as 1 3 = 0.33333333…, which continue indefinitely. We normally disregard decimals that include repeated zeros because they are considered terminating decimals. 0.3, for example, may be represented as 0.300000000…, but it isn't considered a recurring decimal.
- Irrational numbers, such as = 3.14159265…, that go on indefinitely with no discernible pattern.

Integers

- Under this term, the following figures are explained:
- Natural numbers
- Whole numbers
- Integers
- Rational numbers
- Irrational numbers
- Actual figures

Natural Numbers (Numbers) Natural numbers are denoted by the letter N. Counting numbers is another term for them. N is an infinite set of numbers; here are a few of them: N = 1, 2, 3, 4, 5, and so on. Numbers in their entirety W is the sign for whole numbers. This set is identical to N except that it contains zero. W is an endless set of numbers; here are a few of them: W = 0 to 1, 2, 3, 4, 5, and so on.

Integers are denoted by the letters Z or I. It's an infinite set of numbers; here are a few of them: I =..., 3, 2, 1, 0, 1, 2, 3,...; I =..., 3, 2, 1; I =..., 3; I =..., 3; I =..., 3; I =..., 3; I =...,

Numbers That Make Sense the rational numbers are represented by the letter Q, which comes from the word quotient, which signifies the result of dividing one number by another. A rational number is a quotient of two numbers having a denominator that is not zero. Q is a set of infinite numbers. It not only contains all the integers, but also all the positive and negative fractions. Here are some examples of rational numbers: Q =..., 4, 12 3, 2, 7 8, 0, 1, 2, 3, 1 4, 100

Irrational Numbers Irrational numbers are represented by the symbol Q, which stands for the complement of Q and refers to all numbers that are not rational yet are real. Here are some examples of this infinite collection of irrational numbers: , as well as surds like 2, 7, 10, 34.

Real Numbers the real numbers are represented by the letter R, and they are an endless set of numbers. This is the collective designation for all numbers, rational and irrational. The whole numbers are subdivided into natural numbers. The integers are a subset of the whole numbers. The rational numbers are a subset of the integers. Real numbers are divided into two categories: rational and irrational numbers.

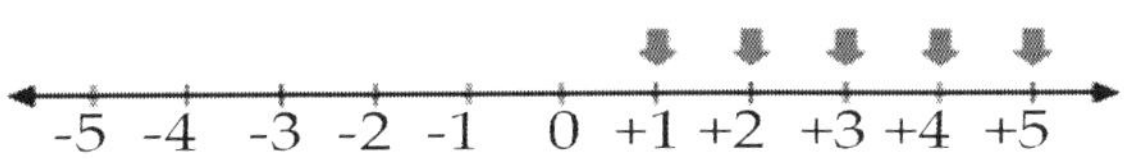

Figure 78. Integers.

Source: Image by Assignment Point.

Intercept

A point on the x-axis or y-axis where the graph of a straight line or a curve intersects the axis is called an intercept. The value of the coordinate at that point can also be an intercept. More information can be found in the references, particularly in the Graphs section.

Interior Angles

Interior and exterior angles exist in polygons. When we talk about a polygon's angles, we usually refer to the interior angles. A pentagon has five sides and five angles, as indicated in the illustration. Because they are all inside the polygon, the five shaded angles are all interior angles. The sum of the pentagon's five internal angles is 540.

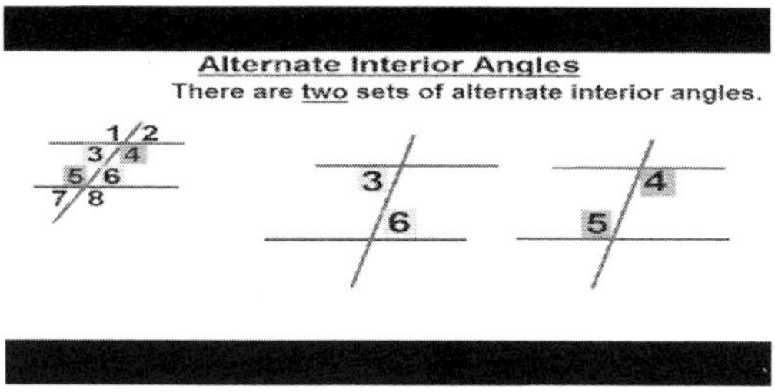

Figure 79. Interior angles.

Source: Image by Wikipedia.

Interquartile Range

We'll look at the median, quartiles (upper and lower), interquartile range, and range in this entry. To explain the various concepts, one set of data is used. The data we utilize is Nathan's class of 25 pupils' test scores on a 10-point scale in a math's test. The item Cumulative Frequency Graph describes the procedures for dealing with enormous amounts of data. The set of 25 marks in the math test, ranked in order of size, is 1, 1, 2, 2, 2, 2, 3, 3, 3, 4, 4, 4, 5, 5, 5, 6, 6, 6, 7, 7, 7, 8, 8, 8, 9.

Intersecting Chords

Under this subject, there are three geometry theorems about the lengths of chords that overlap each other under the following conditions:

- Inside a circle.
- Outside a circle.
- When one of the chords is perpendicular to the circle, it's called a tangent chord.

Theorem 1: When two chords AB and CD cross inside a circle at point X, the chord lengths are related as AX XB = CX XD.

Theorem 2: The lengths of two chords AB and CD intersect outside a circle at a point X (the lines AX and CX are termed secants), are related as AX XB = CX XD.

Theorem no. 3 The lengths of a secant and a tangent to the circle are connected as AX XB = XC2 in the exceptional case of chords that cross outside the circle and one of the chords is a tangent to the circle

Interval

The set of values between two end points of a line is known as an interval. It is a closed interval if the interval includes the end points; it is an open interval if the interval does not include the end points. The collection of all numbers between 3 and 7, including 3 and 7. The solid circles at each end of the interval denote the inclusion of those numbers. If x is the variable, this interval can be represented as 3 x 7 or [3, 7]. This is the end of the interval.

Invariant Points

These are the points that do not change as a result of a transformation. They can be thought of as fixed spots in the face of a transition.

Inverse Operation

An operation's inverse is another operation that "undoes" the original one.

Inverse Relations

Assume we have the relationship "is older than" between three family members: Tom, the grandfather, Joanne, and Luke, the grandson. R = (Tom, Joanne), (Joanne, Luke), (Tom, Luke) represents the set of ordered pairings for this relation.

Investment

An investment is when you lend money to someone or a company like a bank in order to profit. Jo, for example, puts $1,000 in a bank. She makes a 10% profit on her investment after a year: 10% of 1000 = 100, therefore she makes a profit of $100 on her investment.

Isometric

On occasion, we must draw three-dimensional shapes on two-dimensional paper while maintaining their three-dimensional appearance. We can do this by using isometric paper, which has a consistent pattern of dots or spots. The vertices of equilateral triangles are marked with these dots.

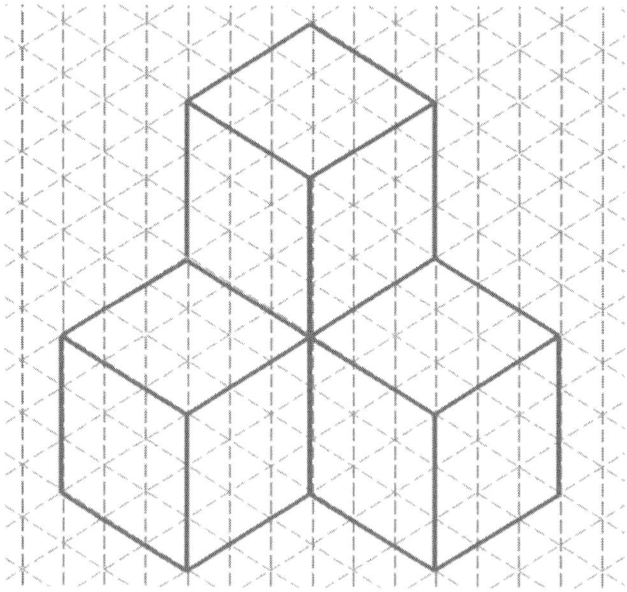

Figure 80. Isometric.

Source: Image by Flickr.

Isosceles Trapezium

This trapezium has two equal-length sides and two sets of congruent angles. It has one axis of symmetry and order one rotational symmetry. When an isosceles triangle is taken from a bigger isosceles triangle, the result is an isosceles trapezium, as seen in the figure. This removing principle applies to equilateral triangles as well. The trapezium is shown in the diagram.

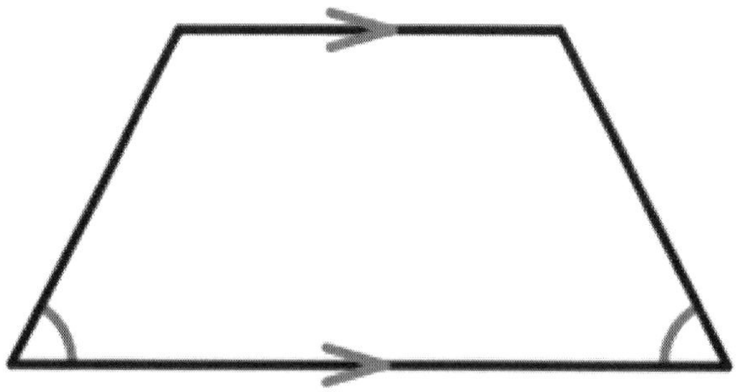

Figure 81. Isosceles trapezium.

Source: Image by Wikimedia Commons.

Isosceles Triangle

A triangle with two equal sides and two equal angles is known as a right triangle. The isosceles triangle's two equal angles are known as the base angles. A geometry theorem is stated as: Theorem. An isosceles triangle has equal base angles. Isosceles triangles all have one axis of symmetry and order one rotational symmetry.

K

Kilo

Kilo is a 1000-unit prefix. One kilometer, for example, equals 1000 meters. Kilometer is abbreviated as km, with k standing for kilo and m for meter. The metric units are primarily based on multiplying a quantity by 1000 (thus the prefix kilo) or dividing a quantity by 1000 (hence the prefix kilogram).

Kilogram

1000 grams equals one kilogram. Kilogram is abbreviated as kg, with k denoting kilo and g denoting gram.

Figure 82. Kilogram.

Source: Image by Earth.com.

Kiloliter

A kiloliter is equal to 1000 liters. The abbreviation for kiloliter is kl, where k stands for kilo and l represents for liter.

Kite

A kite is a quadrilateral with a convex shape. It is convex if all of its angles are fewer than 180 degrees, and quadrilateral if it has four sides. One axis of symmetry runs through a kite, and two pairs of neighboring sides are of identical length. Some kites are shaped like kites and are flown at the end of a string on a windy day.

Konigsberg Bridge Problem

This is a problem from the past that is still relevant today. Leonhard Euler solved an issue that piqued the interest of the citizens of Konigsberg, Prussia, in 1736. The settlement lay near the Pregel River, which split and ran around an island. As seen in the diagram, there were seven bridges spanning the river. Shades have been added to the island.

L

Length

The shortest distance between the end points of a line segment, which is part of a straight line, is its length. A line has only one dimension, but area is a quantity that includes both length and width, with length usually being greater than width. The most common unit for measuring length is the meter.

A ruler or tape measure is used to measure the length of a straight-line segment. A curved line's length can be determined by placing a piece of string along it and measuring the length with a ruler or a flexible tape measure.

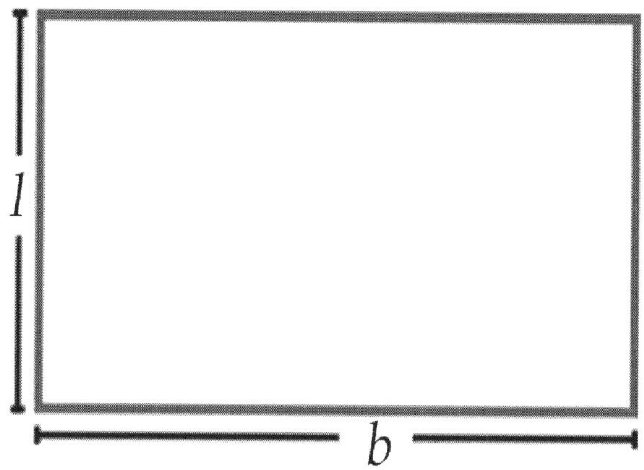

Figure 83. Length.

Source: Image by Splash Learn.

Line

The collection of points that move directly between A and B is represented by the line segment between A and B, which is the first line in the diagram. AB is the abbreviation for the line segment AB. The ray AB, the second line in the diagram, is the infinite continuation in the direction of B of the straight-line segment between A and B. It's abbreviated as AB. The third line in the diagram is the endless continuation of the straight-line segment between A and B in both directions. It is written as AB ↔.

Line Graph

This entry is only for statistical line graphs, and it is only used when the data being graphed is continuous. The example below demonstrates how to create and use a statistical line graph.

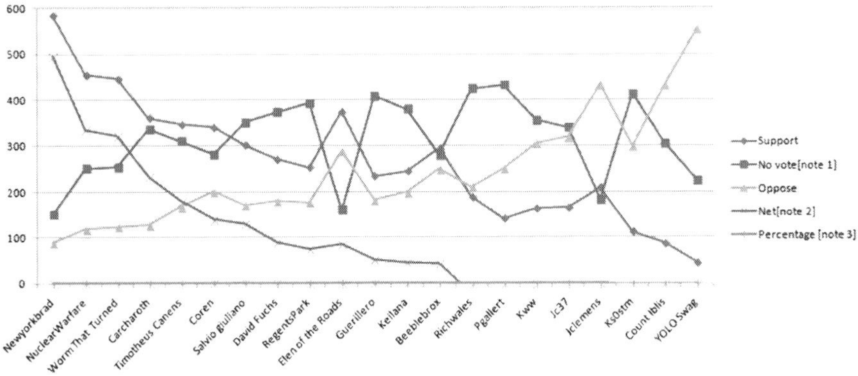

Figure 84. Line graph.

Source: Image by Wikipedia.

Line Segment

In geometry, a line segment is bounded by two distinct points on a line. Or we can say a line segment is part of the line that connects two points. A line has no endpoints and extends infinitely in both the direction but a line segment has two fixed or definite endpoints. The difference between a Line segment and a ray is that a ray has only one endpoint and its other end extends infinitely.

A one-dimensional shape and collection of points extending infinitely in either direction is a line. A part of a line that has two endpoints and is the shortest distance between them, is a line segment.

A ray begins at one endpoint and ends at infinity.

A line segment has two definite endpoints in a line. The length of the line segment is fixed, which is the distance between two fixed points. The length here can be measured by metric units such as centimeters (cm), millimeters (mm) or by conventional units like feet or inches. While learning the concept of lines and angles in Class 7, you will come across the basic terminologies related to it. But in Class 9 the advanced version will be described.

A closed line segment consists of both endpoints, whereas an open line segment is exclusive of the two endpoints. A line segment that has exactly one endpoint is called a half-open line segment.

Lowest Common Multiple

The abbreviation LCM stands for Least Common Multiple. The least common multiple of a number is the smallest number that is the product of two or more numbers. The least common multiple can be calculated for two or more integers as well as two or more fractions. The least common multiple of two numbers is the lowest possible number that can be divisible by both numbers.

There is more than one method to find the LCM of two numbers. One of the quickest ways to find the LCM of two numbers is to use the prime factorization of each number and then the product of the least powers of the common prime factors will be the LCM of those numbers.

The least common multiple is also known as LCM (or) the Lowest Common Multiple in Math. The least common multiple of two or more numbers is the smallest number among all common multiples of the given numbers. Let's take two numbers: say, 2 and 5. Each will have its own set of multiples.

- Multiples of 2 are 2, 4, 6, 8, 10, 12, 14, 16, 18, 20, …
- Multiples of 5 are 5, 10, 15, 20, …

Linear Equation

This is a one-dimensional equation with only one solution. A fundamental linear equation is written as $ax + b = c$, with the variables a, b, and c. Before continuing, you may want to review the entries Balancing an Equation and Inverse Operation. The following illustration of how an equation is constructed provides a practical introduction to solving a linear equation.

Linear Graph

A straight-line graph is known as a linear graph.

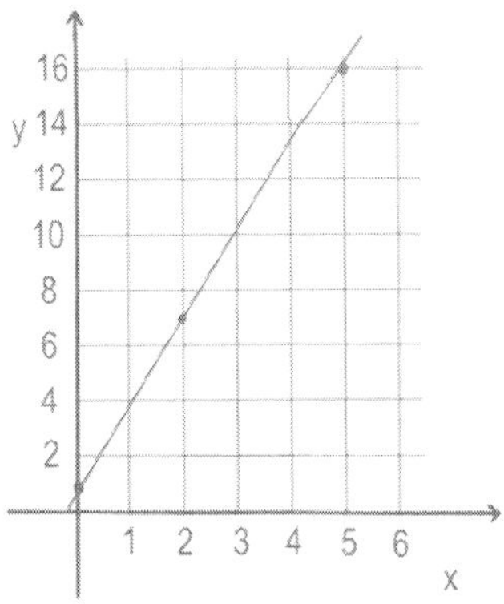

Figure 85. Linear graph.

Source: Image by Wikimedia commons.

Liter

The liter (abbreviated as l) is a unit of volume used to express liquid volume and capacity in general. Milk, for example, is frequently marketed in 1-liter quantities in nations that utilize the metric system. One liter contains 1000 milliliters (abbreviated ml). One liter is equal to a cube with sides of 10 cm. This equates to 1000 cubic centimeters per liter (cm3).

The following are some other useful equivalent relationships: 1 cubic centimeter Equals 1 milliliter 1 cubic meter equals 1000 liters. If the water is at 4 degrees Celsius, a volume of 1000 liters has a mass of one tons. One millimeter of rain equals one liter of water per square meter.

The plural form of locus is loci. A locus is a set of points that form a path, and the positions of each point in the path follow a set of rules. The following example exemplifies this principle. Amanda is preparing George, her horse, for a competition. She stands in one spot, at point A, clutching the end of a rope, which is connected to George on the other end (G).

Amanda spins so that she is constantly facing George as George trots around her, but she keeps the rope taut and the same length. George's various positions make up the set of points, and George's route is the hub of G. The locus of G in this case is a circle. Each point G in this locus is the same distance from the fixed-point A, according to the rule. This locus consists of an infinite number of George positions.

Line Segment

Considering geometry, a line segment can be defined as a line which is bounded by two distinct points. In other words, it can be said that a line segment is part of the line which connects two points. A line has no endpoints and furthermore, extends infinitely in both the direction but a line segment has two fixed or definite endpoints. There can be difference in between a Line segment and a ray and that is that a ray has only one endpoint and its other end extends infinitely.

A line can be defined as one-dimensional shape and collection of points which extending infinitely in either direction. A part of a line which has two endpoints and also, is the shortest distance between them, is referred to as a line segment.

A ray is a line which begins at one endpoint and ends at infinity.

A line segment has two definite endpoints. The length of the line segment is fixed, and it is the distance between two fixed points. In addition to this, the length can be measured by metric units which includes centimeters (cm), millimeters (mm) or even by conventional units such as feet or inches.

A closed line segment contains both end points, while on the other hand, an open line segment is exclusive of the two endpoints. A line segment having exactly one endpoint is referred to a half-open line segment.

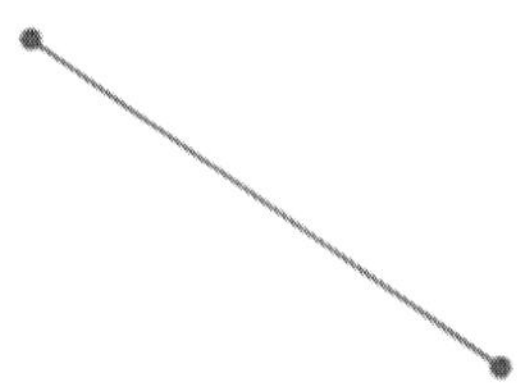

Figure 86. Line segment.

Source: Image by Wikimedia commons.

Lowest Common Denominator

The LCD that is Least Common Denominator is also referred to as the lowest common denominator. It can be defined as the least common multiple for all the denominators in a group of fractions or rational expressions. Considering a group of numbers, the least common multiple is said to be the smallest number which will divide evenly into all of the numbers in the group.

The rational expressions are just fractions, which have variables in their denominators. Let's consider an example, the pizza order that is placed was similar to an LCD in that the order was the smallest number of pizzas which allowed each person to have a sufficient amount of pizza with the toppings he or she desired.

There is a major reason to find the LCD in math and that is when there is a problem where it is required to add or subtract fractions or rational expressions. Along with this, the numerators of fractions can only be combined when the fractions have identical denominators, then finding the LCD is something which requires to be done quite frequently.

Lowest Common Multiple

The LCM is an abbreviation which stands for Least Common Multiple. The least common multiple of a number can be defined as the smallest number which is the product of two or more numbers. The least common multiple can be calculated for either two or more integers and two or more fractions. The least common multiple of two numbers can be defined as the lowest possible number that can be divisible by both numbers.

There is more than one method which can help in finding the LCM of two numbers. Among them, one of the quickest ways to find the LCM of two numbers is using the prime factorization of each and every number and then the product of the least powers of the common prime factors will be the LCM of those numbers.

Also, the least common multiple is referred to as LCM or Lowest Common Multiple in Mathematics. The least common multiple of two or more numbers is said to be the smallest number among all common multiples of the numbers given. Let's consider two numbers as example, say, 2 and 5. Each will have its own set of multiples.

- Multiples of 2 are 2, 4, 6, 8, 10, 12, 14, 16, 18, 20, …
- Multiples of 5 are 5, 10, 15, 20, …

M

Magic Square

A magic square is a square grid of counting numbers in which the sum of the numbers in each row, column, or diagonal in the grid equals the same total. We get the same answer if we add up the numbers in each row, column, or diagonal, which is 34. Because it is a 4 by 4 magic square, the numbers utilized in the grid are 1, 2, 3, 4, 5, 6, 7, 8, 9, 10, 11, 12, 13, 14, 15, and 16.

Figure 87. Magic square.

Source: Image by Wikimedia commons.

Maximum Value

Assume you're driving a car, and your speed is increasing to a high speed at point A , until you get to a bend. Looking ahead, you notice the turn and begin to slow down from your current pace. At point B, your speed slows to a minimum as you proceed around the corner. You accelerate out of the turn, gaining speed along the way. At point C, you're moving quickly once more. A map of the corner is shown in the picture, and a velocity–time graph shows how your car speed changes as you move around the corner. The three points A, B, and C on the map correspond to three points on the graph.

Magnitude of A Vector

The vector is an object having both the magnitude as well as direction. In order to find the magnitude of a vector, it is important to calculate the length of the vector. The quantities including velocity, displacement, momentum, force etc. are vector quantities. But on the other hand, speed, distance, volume, mass, temperature, etc. are scalar quantities. The scalar quantities have the only magnitude, while on the other hand the vectors have both magnitude as well as direction.

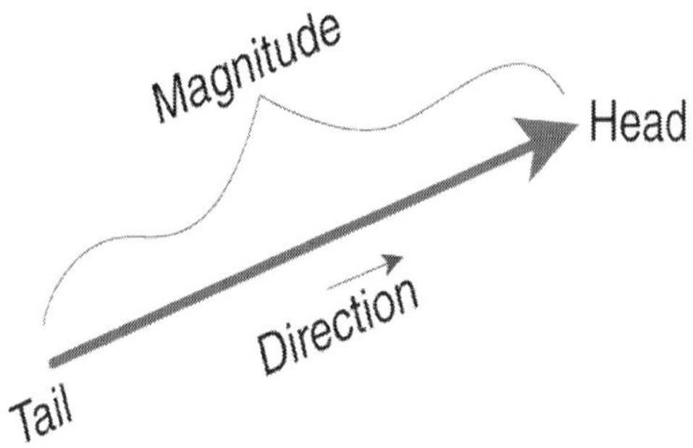

Figure 88. Magnitude of a vector.

Source: Image by Flickr.

The magnitude of a vector formula is used for calculating the length for a given vector (like **v**) and is denoted as $|\mathbf{v}|$. Therefore, typically, this quantity is said to be the length between the initial point and endpoint of the vector.

Mapping

Mapping, any prescribed way of assigning to each object in one set a particular object in another (or the same) set. Mapping applies to any set: a collection of objects, such as all whole numbers, all the points on a line, or all those inside a circle. For example, "multiply by two" defines a mapping of the set of all whole numbers onto the set of even numbers. A rotation is a map of a plane or of all of space into itself. In mathematics, the words mapping, map, and transformation tend to be used interchangeably.

The mathematical notion of mapping is an abstraction of the process of making a geographical map. It is now considered to be a fundamental notion pervading much of mathematics. Important special classes of mappings are homomorphisms in algebra, isometries in geometry, operators in analysis, homeomorphisms in topology, representations in group theory, and isomorphisms in a variety of contexts.

Mean Square Deviation

In statistics, the mean squared error (MSE) or mean squared deviation (MSD) of an estimator (of a procedure for estimating an unobserved quantity) measures the average of the squares of the errors—that is, the average squared difference between the estimated values and the actual value. MSE is a risk function, corresponding to the expected value of the squared error loss. The fact that MSE is almost always strictly positive (and not zero) is because of randomness or because the estimator does not account for information that could produce a more accurate estimate.

The MSE is a measure of the quality of an estimator. As it is derived from the square of Euclidean distance, it is always a positive value with the error decreasing as the error approaches zero.[clarification needed]

The MSE is the second moment (about the origin) of the error, and thus incorporates both the variance of the estimator (how widely spread the estimates are from one data sample to another) and its bias (how far off the average estimated value is from the true value).[citation needed] For an unbiased estimator, the MSE is the variance of the estimator. Like the variance, MSE has the same units of measurement as the square of the quantity being estimated. In an analogy to standard deviation, taking the square root of MSE yields the root-mean-square error or root-mean-square deviation (RMSE or RMSD), which has the same units as the quantity being estimated; for an unbiased estimator, the RMSE is the square root of the variance, known as the standard error.

Median of A Set of Data

Median, in statistics, is the middle value of the given list of data, when arranged in an order. The arrangement of data or observations can be done either in ascending order or descending order.

Example: The median of 2,3,4 is 3. In Math's, the median is also a type of average, which is used to find the center value. Therefore, it is also called measure of central tendency.

Apart from the median, the other two central tendencies are mean and mode. Mean is the ratio of sum of all observations and total number of observations. Mode is the value in the given data-set, repeated most of the time.

In geometry, a median is also defined as the center point of a polygon. For example, the median of a triangle is the line segment joining the vertex of triangle to the center of the opposite sides. Therefore, a median bisects the sides of triangle.

The median of a set of data is the middlemost number or center value in the set. The median is also the number that is halfway into the set.

To find the median, the data should be arranged, first, in order of least to greatest or greatest to the least value. A median is a number that is separated by the higher half of a data sample, a population or a probability distribution, from the lower half. The median is different for different types of distribution.

Mixed Number

Mixed Fractions are one of the three types of fractions. It is also called mixed numbers. For example, $2^1/_7$ is a mixed number.

It is a form of a fraction which is defined as the ones having a fraction and a whole number.

How to convert Improper fraction to a mixed fraction?

- Step 1: Divide the Fraction's numerator with the denominator, i.e., 15/7.
- Step 2: The integer part of the answer will be the integer part for a mixed fraction, i.e., 2 is an integer.
- Step 3: The Denominator will be the same as original, i.e., 7.
- Step 4: So, the improper fraction 15/7 is changed to a Mixed fraction as 2 (1/7)

Figure 89. Mixed number.

Source: Image by Public Domain pictures.

Mode

The mode is the value that appears most frequently in a data set. A set of data may have one mode, more than one mode, or no mode at all. Other popular measures of central tendency include the mean, or the average of a set, and the median, the middle value in a set.

In statistics, data can be distributed in various ways. The most often cited distribution is the classic normal (bell-curve) distribution. In this, and some other distributions, the mean (average) value falls at the mid-point, which is also the peak frequency of observed values. For such a distribution, the mean, median, and mode are all the same value. This means that this value is the average value, the middle value, also the mode—the most frequently occurring value in the data.

Mode is most useful as a measure of central tendency when examining categorical data, such as models of cars or flavors of soda, for which a mathematical average median value based on ordering cannot be calculated.

Mathematical Modeling

Mathematical modeling is the art of translating problems from an application area into tractable mathematical formulations whose theoretical and numerical analysis provides insight, answers, and guidance useful for the originating application.

Mathematical modeling:

- is indispensable in many applications.
- is successful in many further applications.
- gives precision and direction for problem solution.
- enables a thorough understanding of the system modeled.
- prepares the way for better design or control of a system.
- allows the efficient use of modern computing capabilities.

Learning about mathematical modeling is an important step from a theoretical mathematical training to an application-oriented mathematical expertise, and makes the student fit for mastering the challenges of our modern technological culture.

Mapping

Mapping can be defined as a prescribed way of assigning to each object in one set a specific object in another (or the same) set. The implementation of mapping can be done to any set including a collection of objects, like all whole numbers, all the points on a line, or all those that are present inside a circle. Let's consider an example, "multiply by two" which defines a mapping of the set of all whole numbers onto the set of even numbers. A rotation is considered to be a map of a plane or of all of space into itself. In the field of mathematics, the words mapping, map, and transformation is used interchangeably.

The mathematical notion of mapping is an abstraction of the process in order to make a geographical map. In present scenario, it is considered to be a fundamental notion which pervades much of mathematics. The important as well as special classes of mappings are homomorphisms in algebra, homeomorphisms in topology, isometries in geometry, operators in analysis, representations in group theory, and isomorphisms in a variety of contexts.

Mean Square Deviation

Considering the field of statistics, it can be said that the MSE that is mean squared error or MSD that is mean squared deviation of an estimator (of a procedure for estimating an unobserved quantity) tends to measure the average of the squares of the errors which is, the average squared difference between the estimated values and the actual value. MSE is considered to be a risk function, which is corresponding to the expected value of the squared error loss. The MSE is almost always strictly positive (and not zero) and this is said to be true due to the randomness or because the estimator does not account for information that could produce a more accurate estimate.

The MSE can be defined as a measure of the quality of an estimator. Along with the process in which it is derived from the square of Euclidean distance, it is always a positive value having the error decreasing as the error approaches zero.

The MSE is considered to be the second moment (about the origin) of the error, and therefore, incorporates both the variance of the estimator as well as its bias. Considering an unbiased estimator, the MSE is said to be the variance of the estimator. Similar to the variance, MSE has the same units of measurement as the square of the quantity that is being estimated. In an analogy to standard deviation, taking the square root of MSE yields the RMSE that is root – mean – square error or RMSD that is root – mean – square deviation having the same units as the quantity being estimated; for an unbiased estimator, the RMSE is the square root of the variance, referred to as the standard error.

Median of a Set of Data

In statistics, the median is referred to as the middle value of the provided list of data, when arranged in an order. The data or observations can be arranged either in ascending order or descending order.

For example: The median of 2,3,4 is 3.

In Mathematics, the median is also referred to as a type of average, that is used in order to find the center value. Thus, it can also be referred to as measure of central tendency.

In addition to the median, the other two central tendencies are mean and mode. Mean is said to be the ratio of sum of all observations and total number of observations. Mode is said to be the value in the given data – set that is being repeated often.

Considering the field of geometry, a median can also be defined as the center point of a polygon. Let's consider an example, the median of a triangle is the line segment joining the vertex of triangle to the center of the opposite sides. Thus, a median bisects the sides of triangle.

The median of a set of data is referred to as either the middlemost number or center value in the set. Along with this, the median is the number which is halfway into the set. In order to find the median, the data is re quired to be arranged. Firstly, in order of least to greatest or greatest to the least value. In addition to this, a median is said to be a number that is separated by the higher half of a data sample, a population or a probability distribution, from the lower half. The median is different for different types of distribution.

Mean

A scientific calculator can be used to find the mean of a set of data. Refer to the handbook for instructions because calculators range from one manufacturer to the next and may require different instructions.

Measurement

The size or quantity of anything is defined by a measurement. Here are a few examples of measures and the units they are expressed in: A automobile weighs about 850 kilos. A garden has a length of 40 meters. A farm is 2450 hectares in size. My lunch break lasts for 50 minutes. The swimming pool is 21 degrees Celsius. The milk jug has an 800 – milliliter capacity.

Figure 90. Measurement.

Source: Image by Pixabay.

Median of a Triangle

The median of a triangle is the line drawn from one of the triangle's vertex to the midway of the opposite side. The three medians of a triangle cross at the centroid of the triangle, which is labeled G in the diagram. The triangle's medians are divided by the centroid G in a 2:1 ratio.

The line VM in the drawing represents the mediator of the triangle VAB, where M is the midpoint of the opposite vertex V's side AB. One of the qualities of a triangle's median is that it divides the triangle VAB into two equal-area triangles VAM and VBM. The median would be a good way to split a triangle into two "equal" sections using a straight line if you were requested to do so.

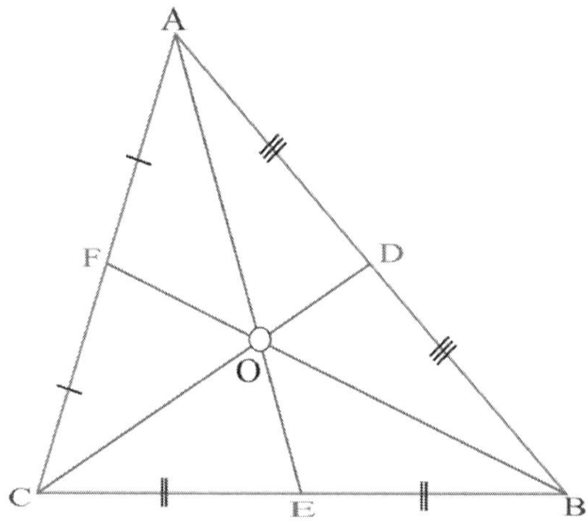

Figure 91. Median of a triangle.

Source: Image by Pixabay.

Mediator

The line that bisects a line segment at right angles is called the mediator. The line that is designated the axis of symmetry of the line segment AB is also the mediator of the line segment AB. The mediator is also known as the line segment AB's perpendicular bisector.

Meter

The abbreviation of meter is m. In the Système International d'Unites, abbreviated SI units, the meter is the base measure of length. The length of the path traveled by light in a vacuum over a time interval of 1/299,792,458 of a second was defined as the standard for the length of 1 meter in 1983. To four decimal places, the lengths comparable to one meter in the imperial system of measurements are 39.3700 inches, 3.2808 feet, and 1.0936 yards. The lengths of races in athletic events are measured in meters, as are the distances thrown and jumped in field events.

Metric Units

The metric unit of length is the meter (m), the metric unit of capacity is the liter (l), and the metric unit of mass is the gram (g) (mass is often incorrectly called weight). Other metric units related to meter, liter, and gram are formed using the prefixes milli (division by 1000) and kilo (multiplication by 1000). Other prefixes exist, and one of the most common is centi (division by 100). Now we'll look at the correlations between various length, area, capacity, volume, and mass units.

Midordinate Rule

This is a formula for calculating the area between a curve and the x-axis.

Mile

The mile is an imperial unit of measurement for length. A mile is around 1.609 kilometers (or 1609 meters) long. A rough guide for rapid conversions between the two units is that 5 miles equals around 8 kilometers.

Milli

This is a prefix that refers to a one-thousandth (1/1000) of a quantity. A millimeter, for example, is one thousandth of a meter and is abbreviated as mm. The first m in this acronym stands for millimeter, and the second m stands for meter. The metric units are primarily based on multiplying or dividing a quantity by 1000 (prefix milli) or dividing a quantity by 1000 (prefix milli) (prefix kilo).

Milliliter

A milliliter (abbreviated ml) is 1/1000 of a liter. One milliliter is equal to one cubic centimeter (cm3) and is a convenient quantity for measuring medicine. A teaspoon of medicine is approximately 5 milliliters.

Millimeter

A millimeter is one thousandth of a meter and is abbreviated as mm. The first m represents millimeters, and the second m represents meters. Millimeters, rather than centimeters, are the unit of measurement for builders and architects. If you're measuring and cutting wood for a house, the most precise measurement you can get is to the millimeter.

Minute

The unit minute can be used in two ways: 1. A minute is 1/60 of a degree, which is a very small angle. To 3 dp, 1 minute = 0.017 degree in decimal form. 1 is the symbol for one minute, and 2 is the symbol for two minutes. A minute is 1/60 of an hour and is used to measure time. To 3 dp, one minute is 0.017 hour. The minute is then broken into seconds, with one second equaling 1/60 of a minute. This means that one minute is 60 seconds long, one hour is 60 minutes long, and one hour is 3600 seconds long.

Mirror Line

When an object is reflected in a mirror to create an image, the item and image are mirror images of one another. The mirror line is m, the object is flag F, and the image is flag F.

Mixed Number

Mixed Fractions can be defined as one of the three types of fractions. Also, it is referred to as mixed numbers. Let's consider an example, $2^1/_7$ is a mixed number.

It is considered to be a form of a fraction that is defined as the ones having a fraction and a whole number.

There are steps which can be done in order to convert Improper fraction to a mixed fraction which include the steps mentioned below:

- Step 1: Divide the Fraction's numerator with the denominator, for example, 15/7.
- Step 2: The integer part of the answer will be the integer part for a mixed fraction, like 2 is an integer.
- Step 3: The Denominator will be the same as original, that is 7.
- Step 4: So, the improper fraction 15/7 is changed to a Mixed fraction like 2 (1/7)

MKS System

The meter, kilogram, and second were the foundations of this unit system. It evolved into the International System of Units (SI).

Mobius Strip

Figure 92. Mobius strip.

Source: Image by Flickr.

A Mobius strip, also known as a band, is a continuous surface with only one side and one edge. It was named after A. F. Mobius (1790–1868), a German mathematician. Making a Mobius strip is quite straightforward. Take a 30 cm long strip of paper and cut it into a long, thin rectangle. Twist one of the ends through half a turn. Bring the two ends together and glue them together at A to form a join. A Mobius strip is the product of this process.

Color one side of the paper that makes up the Mobius strip using a crayon, and you will see that you have colored all the papers. If you start at point A and travel the length of the strip, your finger will end up back at point A after traveling 30 cm = 60 cm. This demonstrates that the Mobius strip has only one surface, as opposed to the original rectangle's two.

The Mobius strip has only one edge as well. In belt-driven devices, the Mobius strip's property of having only one surface can be used. Only the inside of a typical belt comes into touch with the drive shaft. Only the inside of a typical belt comes into touch with the drive shaft. Because the surface in contact with the driving shaft is twice as long when the belt is in the shape of a Mobius strip, the belt will last longer due to reduced wear.

The two shafts it joins turn in opposite directions due to the twist in the belt. Cutting a Mobius strip in half using a pair of scissors along the dashed centerline results in two Mobius strips that are linked together like a chain.

Mode

The mode is defined as the value that appears often in a data set. A set of data can have one mode either more than one mode or either no mode at all. There are other popular measures of central tendency which include the mean either the average of a set, and the median, while the middle value in a set.

While in statistics, data can be distributed in a number of ways. The most often observed distribution is the classic normal (bell-curve) distribution. However, in this, and some other distributions of mode, the mean (average) value falls at the mid-point, it is also the peak frequency of observed values. For this type of distribution, the mean, median, and mode are all the similar value. Thus, this means that this value is the average value, the middle value, also the mode—the often-occurring value in the data.

However, mode is very useful as a measure of central tendency while examining categorical data, for example models of cars or flavors of soda, and for this a mathematical average median value based on ordering cannot be calculated.

Modulus

Moreover, the modulus of a function, which is also known as the absolute value of a function gives the magnitude and absolute value of a number irrespective of the number being either positive or negative. This is due to the fact that it always gives a non-negative value of any number or any variable.

It is represented as

$y=|x|$, $y=|x|$

or

$f(x)=|x|$ $f(x)=|x|$

where $f:R \rightarrow Rf:R \rightarrow R$ and $x \in Rx \in R$

$|x||x|$ is the modulus of xx, where xx is a non-negative number.

If in case xx is positive then f(x)f(x) will be of the same value xx. If in case xx is negative, then f(x)f(x) will be the magnitude of xx.

To sum up the above lines,

$$f(x) = \begin{cases} x \text{ if } x \geq 0 \\ -x \text{ if } x < 0 \end{cases}$$

Thus, this means that, if the value of xx is greater than or equal to 0, thus the modulus function takes the actual value, while if in case xx is less than 0 then the function takes minus of the actual value 'x.'

Multiplicand

If two numbers, say 6 and 8, are multiplied together, the result is: $6 \times 8 = 48$. The multiplicand (6) is the first number, the multiplier (8) is the second, and the product (48) is the third.

Fraction Multiplying

It's easier to multiply fractions than it is to add and subtract them. We multiply the numerators and denominators together and express the result as a single fraction when we multiply two fractions together.

Mutually Exclusive Events

While in probability theory, there are two events which are said to be mutually exclusive if in case they cannot occur either at the same time or simultaneously. Moreover, in other words, mutually exclusive events are known as disjoint events. If in case two events are considered disjoint events, then the probability of both events occurring at the same time will be zero. If in case A and B are the two events, then the probability of disjoint of event A and B is written by:

Probability of Disjoint (or) Mutually Exclusive Event = P (A and B) = 0

While in probability, there is specific addition rule which is valid when two events are mutually exclusive. It also states that the probability of either event occurring is the sum of probabilities of every event occurring. If in case A and B are said to be mutually exclusive events thus the probability of an event A occurring or the probability of event B occurring is given as P(A) + P(B), i.e.,

P (A or B) = P(A) + P(B)

For some of examples of the mutually exclusive events are:

First, in a six-sided die, the events "2" and "5" are mutually exclusive. It is not possible to get both the events 2 and 5 at the same time when we threw one die.

- Second, when a coin is tossed, the event of getting head and tail is mutually exclusive. Because the probability of getting head and tail simultaneously is always 0.
- Third, while in a deck of 52 cards, drawing a red card and drawing a club are mutually exclusive events because all the clubs are black.

If in case the events A and B are not mutually exclusive, there is probability of getting A or B is given as:

P (A or B) = P(A) + P(B) – P (A and B)

Dependent and Independent Events

Moreover, the two events are said to be dependent only if the occurrence of any one event changes the probability of another event. Thus, two events are said to be independent events if in case the probability of one event that does not affect the probability of another event. If in case two events are mutually exclusive, they are not independent. Therefore also, independent events cannot be mutually exclusive.

Mutually Exclusive Events Probability Rules

While in probability theory, if in case two events are mutually exclusive or disjoint if they do not occur at the same time. While a clear case is the set of results of a single coin toss, which can end in either heads or either tail, but not for both.

Therefore, while tossing the coin, both outcomes are collectively exhaustive, which suggests that at least one of the consequences will happen in any case, so these two possibilities collectively exhaust all the possibilities. Moreover, not all mutually exclusive events are commonly exhaustive.

Let's considers an example, the outcomes 1 and 4 of a six-sided die, when we roll it, are mutually exclusive (both 1 and 4 cannot come as result at the same time) but not collectively exhaustive (it can result to distinct outcomes such as 2,3,5,6).

From the definition of mutually exclusive events, certain rules for the probability are concluded.

Addition Rule: P (A + B) = 1

Subtraction Rule: P (A U B)' = 0

Multiplication Rule: P (A ∩ B) = 0

Thus, there are different varieties of events also. Like for instance, think a coin either has a Head on both the sides of the coin or a Tail on both sides. While it doesn't matter how many times you flip it will always occur Head (for the first coin) and Tail (for the second coin). Thus, if in any case it is checked that the sample space of such experiment, it will be either { H } for the first coin and { T } for the second one. Thus, such events have single point in the sample space and are called "Simple Events." So, these kind of two sample events is always mutually exclusive.

N

Natural Numbers

Natural numbers, which include all positive integers from 1 to infinity, are a part of the number system. Natural numbers, which do not include zero or negative numbers, are also known as counting numbers.

They are only positive integers, not zero, fractions, decimals, or negative numbers, and they are a part of real numbers.

Natural numbers are those that can be counted and are a portion of real numbers. Only positive integers, such as 1, 2, 3, 4, 5, 6, etc., are included in the set of natural numbers.

Negative

This term refers to a value that is less than zero, whereas positive refers to a value that is greater than zero. On the number line, negative numbers are to the left of zero, while positive numbers are to the right of zero. Negative numbers are not to be confused with the word "minus," which means "to subtract one number from another."

Negative Integers

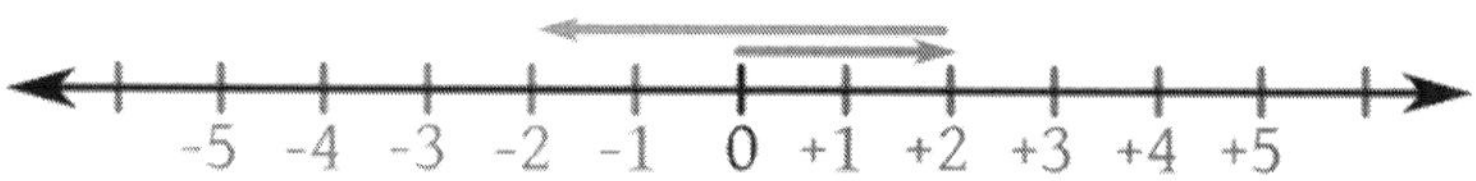

Figure 93. An image showing both negative as well as positive integer.

Source: Image by Wikimedia Commons.

The set of integers consists of{..., −3, −2, −1, 0, 1, 2, 3, ...}. Negative integers are the numbers ..., 3, 2, 1, which are all to the left of zero. Positive integers, on the other hand, are all the numbers to the right of zero, such as 1, 2, 3, and so on. The positive integers and the counting numbers are the same thing. NET is a number that is neither positive nor negative.

The term "net" refers to a two-dimensional drawing that may be cut out and folded to form a solid shape. Pentominoes investigates the many nets of an open cube. The nets of solid shapes such as the cube, cuboid, cylinder, cone, tetrahedron, and others are shown in their own entries.

Networks

The branch of mathematics known as topology includes the study of networks, often known as graph theory. A network is made up of a collection of points connected by lines. Arcs are the lines that connect the points, while nodes are the points. It is not necessary for the lines to be straight. The network's confined spaces are referred to as regions.

A node is a location where two or more arcs intersect or where an arc begins. The amount of arcs leaving a node determines its order. The order of each node is shown by the number on the side of each node. Aside from Euler's formula, another important fact concerning networks is that a network must have an even number of odd nodes in order to exist.

A network is traversable if it can be drawn without removing the pencil from the paper and only going over each arc once. A node can be passed through multiple times, but a line cannot be redrawn. Whether you return to your starting point or not, Euler determined the following rule for a network to be traversable: a network must contain either two odd nodes or no odd nodes to be traversable. It can have as many even nodes as it wants.

Nonagon

Figure 94. Nonagon having nine sides.

Source: Image by Wikimedia Commons.

The nonagon, sometimes known as the enneagon, is a nine-sided polygon with nine sides. Although the term "enneagon" is arguably preferable (since it uses the Greek prefix and suffix rather than the mixed Roman/Greek nonagon), the term "nonagon" is used in this work because it is easier to spell and pronounce.

Nonagons come in a variety of shapes and sizes. A regular polygon known as a regular nonagon is a nonagon with vertices evenly placed around a circle and all sides the same length.

A normal nonagon has nine equal-length sides, and each of its nine angles is the same size and equals 140 degrees. Because there are 360 degrees in a full rotation, the regular nonagon is made up of nine isosceles triangles, and each angle at the nonagon's center is 40.

Normal

To understand what a normal is, you must first understand what a tangent is. A tangent to a curve is a straight line that intersects the curve at a point and has the same gradient (or direction) as the curve. Figure a shows examples of tangents to curves, with a dot indicating the tangent's point of contact with the curve. The illustration on the far right shows a point of inflection for the tangent to the curve. The tangent changes from one side of the curve to the other at a point of inflection.

In connection to a tangent to a curve, we may now explain the word normal to a curve. A straight line that is at right angles to the tangent to the curve and goes through the same point of contact with the curve as the tangent is called a normal to the curve. The tangents are dashed in , and the normals are drawn at right angles to the tangents. The center of a circle is where the circle's normal passes through.

Normals and Tangents Gradients the product of the gradients of a tangent and a normal drawn to a curve at the same position is 1. This rule applies to any two lines that are at right angles in the same plane. If the tangent's gradient is m_1 and the normal's gradient is m_2, then

$$m_1 \times m_2 = -1$$

If two lines with gradients m_1 and m_2 are perpendicular, turning m_1 upside down and changing its sign is a rapid way to get m_2. We've just talked about the normal to a curve so far, but there can also be a normal to a plane. A plane's normal is a straight line that is perpendicular to the plane.

Normal Distribution

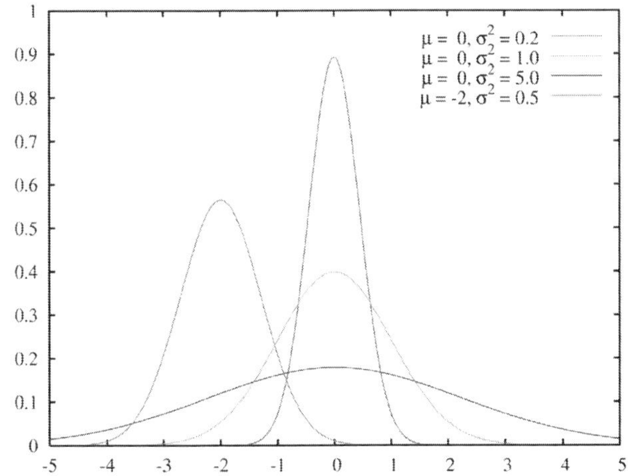

Figure 95. A graph representing normal distribution.

Source: Image by Wikimedia Commons.

The study of probability and statistics includes the study of normal distributions.

The means and standard deviations of normal distributions are not always the same. Even when scores follow a normal distribution, each data set or distribution will have its own mean, standard deviation, and shape.

A normal distribution with a mean of 0 (u=0) and a standard deviation of 1 (o= 1) is known a standard normal distribution or a Z-distribution.

- The standard normal distribution (graph below) is a mathematical- or theoretical distribution that is frequently used by researchers to assess whether the distributions of the variables they are studying approximately follow a normal curve.
- Every score in a normally distributed data set has an equivalent score in the standard normal distribution. This means that the standard normal distribution can be used to calculate the exact percentage of scores between any two points on the normal curve.
- For the typical normal curve, statisticians have devised tables that show the percentage of scores that differ between any two points. Scores must be converted to Z scores before they can be used in this table.

Continuous data refers to data that is collected through measuring, such as people's weights. People's heights, the lengths of leaves on a tree, the life of a light bulb, the amount of paint in a 10-liter container, and examination marks calculated in percentages are all examples of continuous data. Exam results are not continuous data, but they are included here; see the note at the end of this section for further information.

When we graph data using frequency polygons, we are measuring continuous amounts that occur naturally in everyday life.

Number Bases

A number base is the amount of digits or combinations of digits used to represent numbers in a counting system. Any whole number greater than 0 can be used as a basis. The decimal system, sometimes known as base 10, is the most widely used number system. The fact that we have ten fingers is perhaps the reason for its popularity as a counting system.

Number Line

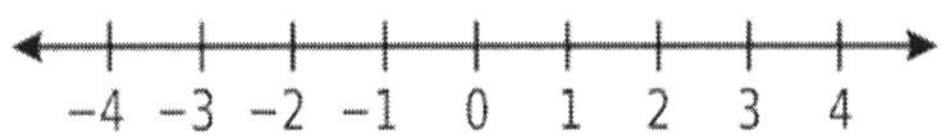

Figure 96. Number line.

Source: Image by Wikimedia Commons.

A number line is a diagram that depicts numbers in a straight line. It serves as a guide to comparing and sorting numbers. Any real number, including whole numbers and natural numbers, can be represented by it. To refresh your memory, the whole number is a collection of numbers that includes all counting numbers (1, 2, 3, 4, 5, 6, …) as well as zero (0), whereas the natural number is the collection of all counting numbers, i.e., 1, 2, 3, 4, 5, 6,….

It's easier to compare numbers when they're written on a number line. The integers on the left side are smaller than the integers on the right side. 0 is less than 1, –1 is less than 0, –2 is less than –1, and so on.

On a number line, arithmetic operations of numbers can be better explained. To begin, one needs understand how to find numbers on a number line. A number line's midway point is zero. Positive numbers occupy the right side of the zero on the number line, while negative numbers occupy the left side of the zero. The value of a number falls as we move to the left side. 1 is greater than –2, for example. Integers, fractions, and decimals can all be simply expressed on a number line. To discover more, click on the links below.

Numerator

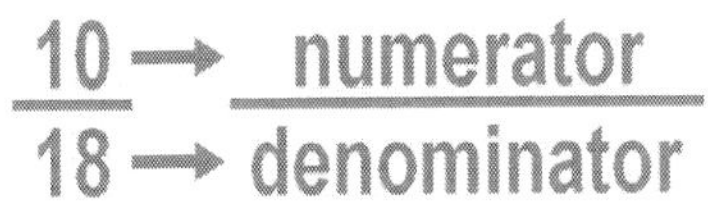

Figure 97. An image showing both numerator as well as denominator.

Source: Image by Flickr.com.

The numerator is the part of a fraction that is on top, while the denominator is the part that is on the bottom.

A fraction's numerator and denominator can be defined in a number of different ways. Keep in mind that a fraction is a portion of a whole. The denominator reflects the total number of parts formed from the whole, whereas the numerator represents how many parts of that whole are being considered. The full value (say, an apple pie) has been divided into 5 equal parts in the fraction above. Three of the five equal halves belong to someone who possesses 3/5 of the pie.

Numerical Value

The value of the phrase in which each of the variables was replaced with a number from its defined domains.

In the case of a single-digit number, talking about the value of a digit is a short form for referring to the numerical value associated with a digit.

Because the constants are all numbers, the equation $3x + 2 = 8$ is a numerical equation. Because the constants are represented by the letters a, b, and c, the equation $ax\ 2 + bx + c = 0$ is known as a literal equation. In place of absolute value, the word numerical value is occasionally used. The numerical value of 7 and +7, for example, is 7.

O

Obtuse Angle

An obtuse angle is one of the numerous types of angles that can be generated on a plane surface. An obtuse angle is one that is more than 90° but less than 180° in Geometry.

In Geometry, an obtuse angle is defined as an angle with a measure of larger than 90° but less than 180°.

In other terms, an obtuse angle is the angle formed by the intersection of a right angle with a straight angle.

Octagon

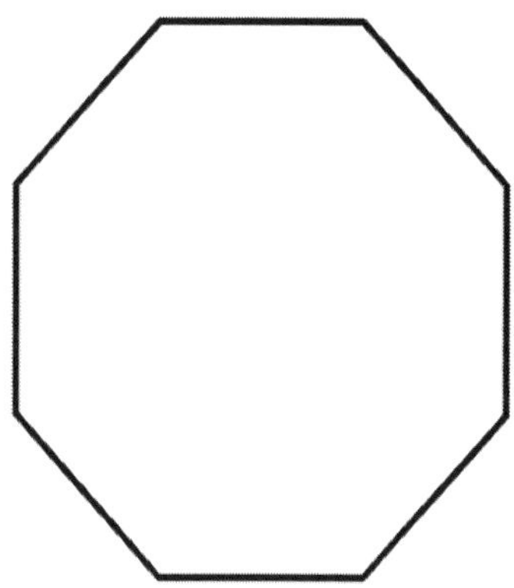

Figure 98. Octagon having eight sides.

Source: Image by Wikimedia Commons.

In geometry, is a polygon with 8 sides and 8 angles. This indicates that there are eight vertices and eight edges. To form a shape, all of the sides are connected end-to-end. These sides are in a straight line with each other; they are not bent or disjointed. A regular octagon has 135° inside angles. As a result, the outside angle is calculated as 180° – 135° = 45°. The sum of the octagon's inner angles is 135 8 = 1080°.

A regular octagon has eight equal-length sides and eight equal-size angles, for a total of 135 angles. The standard octagon is made up of eight isosceles triangles, and each angle at the octagon's center is 45 degrees.

Octahedron

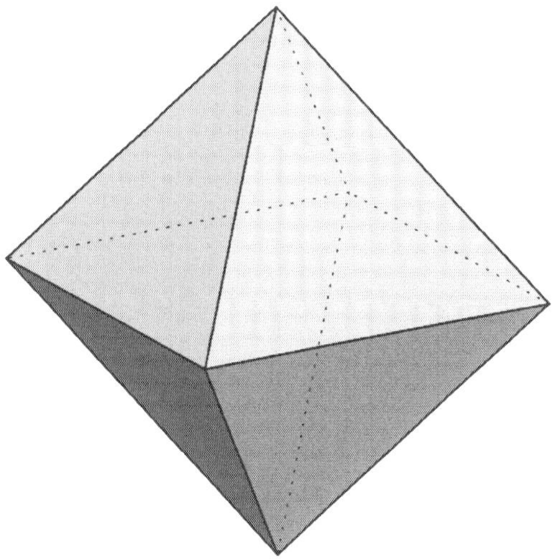

Figure 99. Euclid octahedron.

Source: Image by Wikimedia Commons.

The word octa signifies eight, and this is a solid shape (polyhedron) with eight faces. A regular octahedron, which is one of the Platonic Solids, has eight faces that are all congruent equilateral triangles. The octahedron in the illustration is made up of two square-based pyramids glued together. Eight equilateral triangles make up the net of a normal octahedron.

Odd function

We call a function odd when

$-f(x) = f(-x)$ for all x

Note that the minus in front of f(x): −f(x

They were given the name "odd" because the functions x, x3, x5, x7, and others behave in this way, but other functions, such as sin, also behave in this way (x). It's a peculiar feature.

However, a function with an odd exponent is not always an odd function; for example, x3+1 is not an odd function.

Odd Numbers

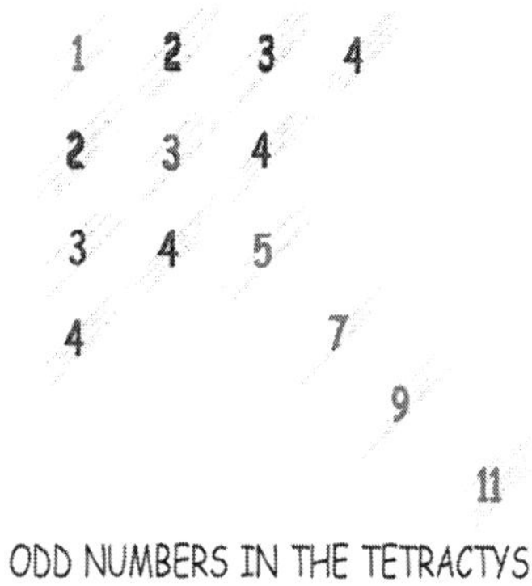

Figure 100. Odd numbers in the tetractys.

Source: Image by Flickr.com.

Numbers that cannot be organized in pairs are known as odd numbers. Numbers that could not be placed in two rows were considered odd by the Ancient Greeks. Over the millennia, this concept has evolved. Take any two-digit multiple as an example. You'll see that none of these numbers can be grouped in two-digit pairs. All integers, with the exception of multiples of 2, are odd numbers. Odd numbers are defined as those numbers that cannot be divided evenly into two pieces.

Odd numbers are simply integer numbers that can't be divided into two groups of two. For instance, 1, 3, 5, 7, and so on. Let's look at it through the lens of footwear and cherries. Assume we have footwear in the following counts: 1, 3, 5, and 7. On the other hand, we have cherries in twos, fours, sixes, and eights. To understand how the pairing of these numbers will function, look at the graphic below.

Odds

It is necessary to read the entry first in order to comprehend the odds. An event's probability. The likelihood of an event occurring is defined as the ratio of the probability of it occurring to the probability of it not occurring. Assume that the chance of an event occurring is P. Then there's a 1–P chance that it won't happen. As a result, the probability of an event occurring is P/(1 – P). By the use of the definition

Ogive

The graph of a cumulative frequency distribution is known as an ogive.

Cumulative histograms, also known as ogives, are graphs that can be used to determine how many data values in a data set are above or below a specific number. The cumulative frequency is computed by adding each frequency to the total of the frequencies of all data values in the data set before it in a frequency table. Because all frequencies have previously been added to the previous total, the last value for the cumulative frequency will always be equal to the entire number of data values.

- An ogive is made by starting the first interval with a (y)-value of zero.
- ending each interval with a (y)-value equal to the cumulative count for that interval.
- and connecting the points on the plot with straight lines.

Operations

A way of merging numbers is called an operation. Addition, subtraction, multiplication, and division are the four operations. You do not need to be able to perform these calculations by hand because Functional Skills exam questions enable the use of a calculator throughout; nonetheless, you must comprehend what each operation accomplishes and when to utilize it. To answer a question at level 2, you will frequently be required to use more than one operation.

(+) Addition

When you wish to find the total, or sum, of two or more amounts, you perform this procedure.

(-) Subtraction

When you need to find the difference between two amounts or how much of something is left after a quantity has been consumed – for example, if you need to find the change owing after spending a certain amount of money – this operation is employed.

(×) Multiplication

This procedure is also used for totals and sums, but only when there are multiples of the same number — for example, if you want to buy five bags of sand for £42 apiece, you'd do 5 £42.

(%) Division

is employed When sharing or grouping objects. For example, if one can of paint costs £14, you would do £84 £14 to find out how many cans you can buy with £84.

Opposite Integers

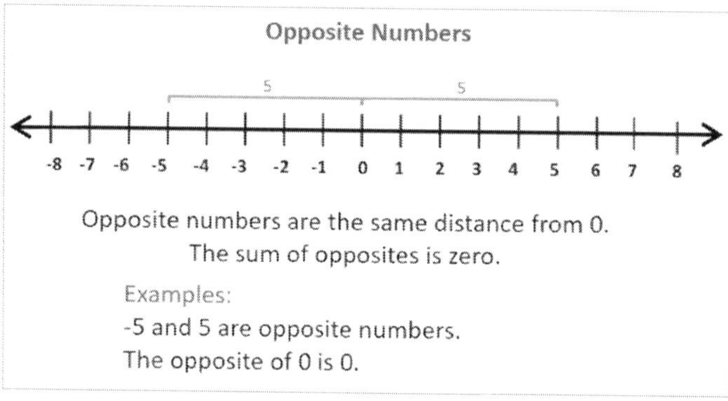

Figure 101. Opposite integers.

Source: Image by Pixabay.

When two integers are placed on a number line on opposite sides of the origin and at equal distances from the origin, one is the opposite of the other. The word "opposite of an integer" refers to the process of subtracting integers, which is covered in the item Integers. For instance, the inverse of 2 is +2, or just 2. 4 is the inverse of 4.

Opposite Transformation

A transformation that changes the orientation of a figure is known as an opposing transformation.

Glide reflections and reflections are the polar opposites of each other.

The original graphic, triangle ABC, for example, has a clockwise orientation, with the letters A, B, and C read clockwise.

The image triangle A'B'C' has a counterclockwise orientation after reflection in the x-axis – the letters A,' B,' and C' are read in a counterclockwise direction. A reflection is an opposite transformation.

Order of Rotational Symmetry

The number of times you can spin a geometric object to make it seem exactly like the original figure is called its order of rotational symmetry. All you have to do is rotate the figure 360 degrees. You'll be returned to the original figure once you've rotated the figure 360 degrees.

Let's start with a geometry with a rotational symmetry order of 1. A rotational symmetry of order 1 means that after rotating the shape 360 degrees, it will only look like it did before. You'll see that until you rotated the arrow 360 degrees, you couldn't get the original arrow.

Total Symmetry Order This is the consequence of multiplying a shape's number of axes of symmetry by its order of rotational symmetry. A square's total order of symmetry, for example, is 4 + 4 = 8. Search for the symmetry of well-known polygons under the relevant entry.

Ordinal Data

Ordinal data is a sort of statistical quantitative data in which variables are organized into naturally occurring groups. Ordinal data is not used to determine the distance between two categories.

In statistics, ordinal data is represented by a group of ordinal numbers, and ordinal data is represented by an ordinal scale. The primary distinction between nominal and ordinal data is that ordinal data has a category order whereas nominal data does not.

A common example of ordinal data is the Likert Scale. A Likert Scale will include the following alternatives coded to 1, 2, 3, 4, and 5 for a question like: "Please indicate the importance of cost for you to acquire a product" (numbers). 1 is lower than 2, which is lower than 3, which is lower than 4, which is lower than 5, and so on.

Ordinal data is a collection of ordinal variables, i.e., if you have variables in a specific order, such as "low, medium, high," you can describe them as ordinal data. For ordinal data, there are two crucial variables to consider: "order" can be expressed in a variety of ways, such as "high, higher, highest" or "satisfied, dissatisfied, extremely dissatisfied."

The disparity between variables isn't even.

- Establish a relative ranking: In the example above, slightly pleasant is clearly worse than extremely pleasant, and very unpleasant is surely worse than somewhat unpleasant. There is a distinct rank inside the alternatives, indicating ordinal data.

- The difference between very nice and somewhat pleasant does not have to be the same as the difference between somewhat disagreeable and very awful. The ordinal scale cannot be used to determine this interval.
- Measure non-numeric traits: All of the answer alternatives in the example are non-numeric, and ordinal data can be used to represent feelings like pleasure, happiness, and frequency.
- Nominal data with an add-on: Nominal data is data that has been "labelled." Ordinal data is data that has been labelled in a specified order. The alternatives in the sample above are in a noticeable order, making it a classic case of ordinal data.
- Ordinal data has a median: The median is the value in the center of a scale that is not the same as the middle value, and it can be calculated using data with inherent order.

Origin

The intersecting point of the axes of the coordinate system. The extent to which every coordinate is 0

In two dimensions

There are two axis, vertical y-axis and horizontal x-axis, in the flat coordinate plane. The source is the crossroads of them. The point is 0,0 and commonly marked with the letter O.

In three dimensions

A third axis, the z-axis, runs in and out of the screen at right angles to the other two in the three-dimensional coordinate system. The point O where all three axes intersect is the origin. The coordinates of this place are 0,0,0.

Orhocenter

The orthocenter is the place where the triangle's three altitudes cross or cut each other. The height is defined as a line drawn from the triangle's vertex that is perpendicular to the opposite side. There are three heights because the triangle has three vertices and three sides.

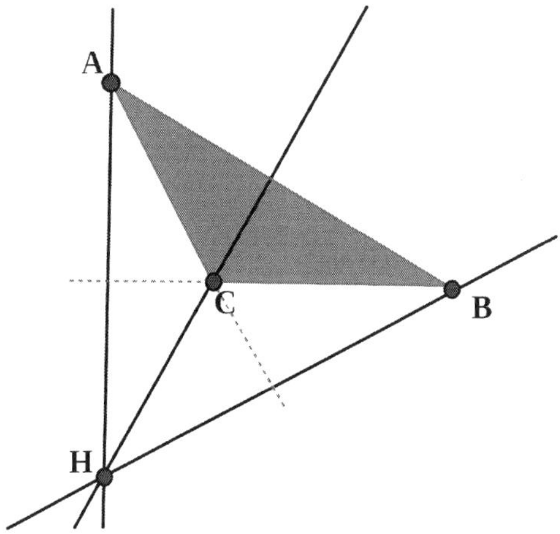

Figure 102. Orthocenter of obtuse triangle.

Source: Image by Wikimedia Commons.

Distinct types of triangles have different orthocenters, such as isosceles, equilateral, scalene, right-angled, and so on. The orthocenter will be the centroid in the case of an equilateral triangle. However, the position of other triangles will be different. The orthocenter does not have to be inside the triangle; in the case of an obtuse triangle, it might be outside.

The point where the perpendicular drawn from the vertices to the opposing sides of the triangle intersect each other is the orthocenter of the triangle.

- The orthocenter of an acute angle triangle is located within the triangle.
- The orthocenter of an obtuse angle triangle is outside the triangle.
- The orthocenter of a right triangle is the vertex of the right angle.

Ounce

The ounce (abbreviated as oz) is a weight unit that is used in imperial measurements. One-ounce equals 1/16 of a pound, so 16 oz = 1 pound (lb). The weight of an ounce is approximately 28 grammes (28.349532 grams, to 6 dp).

Figure 103. Vintage one-ounce fine gold bar.

Source: Image by Flickr.

Outcome

An outcome is the result of a probability-based occurrence. Probability relates to the possibility of anything happening, and an event can have multiple outcomes.

Oval

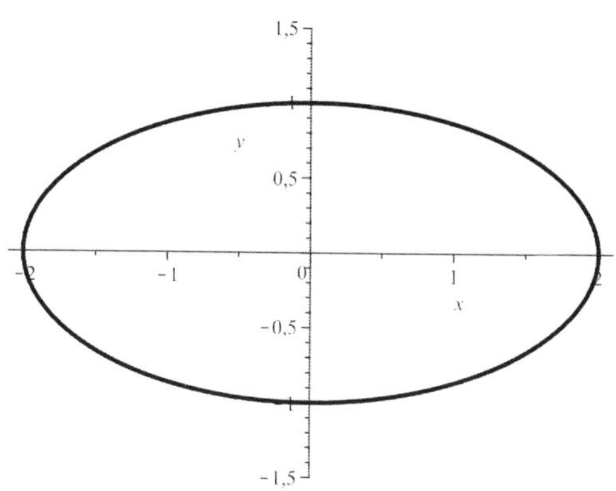

Figure 104. Ellipse equation to oval.

Source: Image by Wikimedia Commons.

An oval is a curve that resembles a squashed circle but lacks a clear mathematical definition, unlike the ellipse. The word oval comes from the Latin word "ovus," which means "egg." In contrast to ellipses, ovals can have just one axis of reflection symmetry (instead of two).

P

Palindromic Numbers

This is a number that reads the same in both directions. The number 1221, for example, is a palindromic number since reading it backwards yields the same number, 1221. 12621, 22, and 4884 are examples of palindromic numbers. A palindromic number is formed by adding the last digits of each of the first nine square numbers:

12 = 1, 22 = 4, 32 = 9, 42 = 16, 52 = 25, 62 = 36, 72 = 49, 82 = 64, 92 = 81.

We get the palindromic number 149656941 by writing down the last digit of each square number.

Parabola

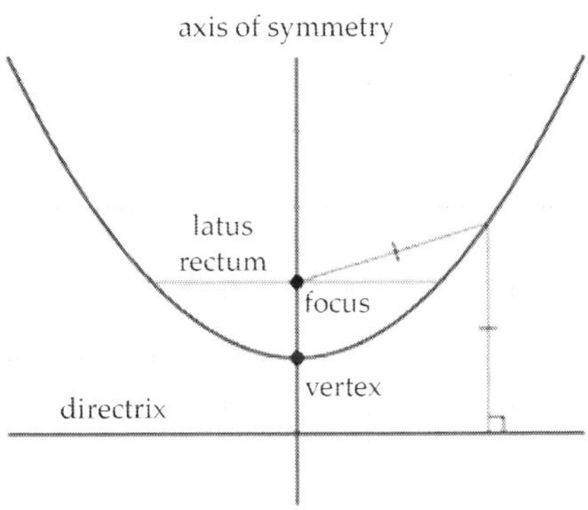

Figure 105. Parts of parabola.

Source: Image by Wikimedia Commons.

For information on how to draw parabolas, see the item Quadratic Graphs. Assume you're tossing a stone across the air. The quadratic formula h = 20t 5t 2 can be used to model its height h meters after time t seconds. The parabola in the illustration is the graph of this quadratic formula. Galileo (1564–1643) was the first to notice that when things are flung through the air, their path has the shape of a parabola.

This is only true when gravity is homogeneous and air resistance is ignored. A parabola is the path of bullets shot from a cannon or bow and arrow, or water sprayed from a hosepipe. Car headlights make use of one of the parabolic curve's features. The light rays emanate from the headlight as parallel beams of light if a light bulb is placed at the focal point (F) of a parabolic reflector. Other curves do not have this attribute.

Another application of the parabolic curve is in astronomy, where it is employed in telescopes and radio telescopes. A parabolic "dish" is designed to collect parallel radio waves from distant planets and stars and focus them at its center.

Paradox

Paradox is a statement that appears to be logical but, upon closer examination, yields an obviously nonsensical result. Zeno, a Greek philosopher who flourished around 450 BC, proposed perhaps the most famous paradox.

Any statement (or collection of statements) that appears to contradict itself (or each other) while yet being entirely logical is referred to as a mathematical paradox. A paradox (at least in mathematics) is simply a false statement that appears correct due to a lack of essential reasoning or information, or the application of logic to a situation where it is inappropriate. In mathematics, there are numerous paradoxes.

Many proofs employ proof by contradiction, in which you make a proposition and then produce a contradiction to show that it is false.

Parallel

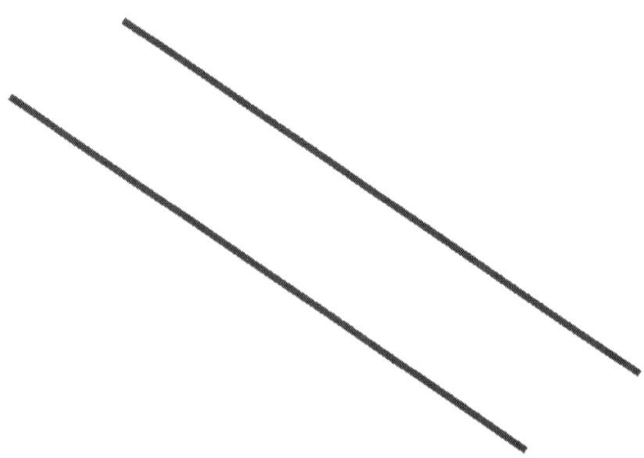

Figure 106. Two parallel lines.

Source: Image by Wikimedia Commons.

If two lines in two-dimensional Euclidean space do not intersect, they are said to be parallel. Parallel lines in three-dimensional Euclidean space not only do not overlap, but also maintain a constant gap between the points nearest to each other on the two lines. Skew lines are lines in three-space that are not parallel but do not intersect.

When lines and are parallel, the notation is used.

If lines and are parallel, the notation is used.

In non-Euclidean geometry, the concept of parallelism must be altered from its intuitive understanding. This is accomplished by modifying the so-called parallel postulate. While the conclusions are unexpected, the geometries provided are completely self-consistent.

Parallelogram

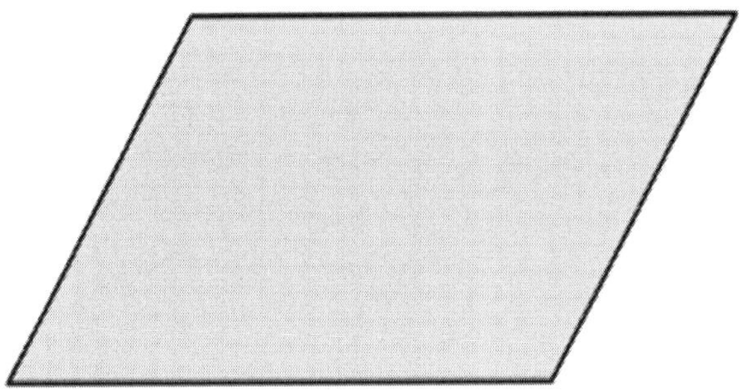

Figure 107. An image showing parallelogram.

Source: Image by Wikimedia Commons.

A parallelogram is a quadrilateral, which is a four-sided polygon with parallel and equal length opposite sides. Furthermore, its opposing angles are of equal size. The angles a and b are known as cointerior angles, and their sum is 180 degrees, or $a + b = 180$.

The parallelogram is divided into two congruent triangles by each diagonal. The parallelogram's diagonals are bisected by each other. It has no symmetry axes. It possesses order two rotational symmetry.

The shapes rhombus, rectangle, and square are all considered special types of parallelograms, although each has its unique set of characteristics. More information about these shapes can be found in their own entries. To calculate the area of a parallelogram, we must first determine the lengths of its base (b) and altitude, or perpendicular height (h). If we only have the lengths of the parallelogram's four sides, we can't calculate its area.

Parentheses

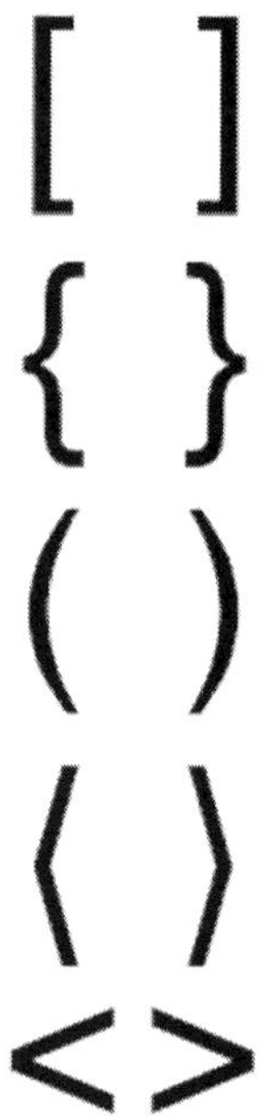

Figure 108. Different types of parentheses.

Source: Image by Wikimedia Commons.

Round brackets ( ) are a type of bracket used in algebra to surround an equation so that it can be handled as a single quantity. When the brackets have served their job, they are frequently expanded and deleted.

Pascal's Triangle

In the entry Coefficient, the utilization of Pascal's triangle and its symmetry patterns are explained. Natural numbers, triangular numbers, and powers of two are all included in Pascal's triangle.

Six rows are drawn in this triangle, but more rows can be added using numerical patterns. The set of counting numbers 1, 2, 3, 4, 5,…, often known as natural numbers, is one of the diagonal rows. A diagonal of triangle numbers follows the natural numbers, 1, 3, 6, 10, and so on. When you add the numbers in each row together, you get powers of 2, which are 1, 4, 9, 16, 32, and so on.

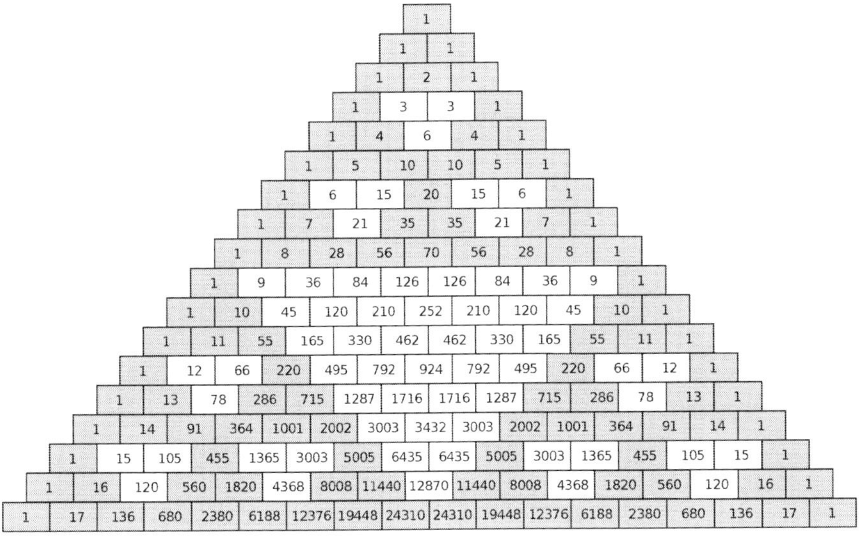

Figure 109. Pascal's Triangle divisible by 3.

Source: Image by Wikimedia Commons.

Patterns

When a set of numbers follows a set of rules, it forms a pattern. A pattern can also be found in a collection of drawings or designs that are represented by a rule. The goal of this post is to discover a rule, or formula, that explains a pattern of numbers or images. The rule will be expressed in terms of n, and the pattern of numbers will be obtained by substituting the numbers n = 1, 2, 3, 4,… into the rule one by one.

Pentagon

A pentagon is a five-sided polygon. Figure a depicts a variety of pentagons, the final of which is a regular pentagon. The regular pentagon has five sides that are all the same length and five angles that are all the same size (108). It contains five symmetry axes and a rotational symmetry order of five. Five congruent isosceles triangles with angles of 54, 54, and 72 make up the regular pentagon. You can use a protractor to measure the 108 angles to make a regular pentagon. Tessellation is not possible with a regular pentagon.

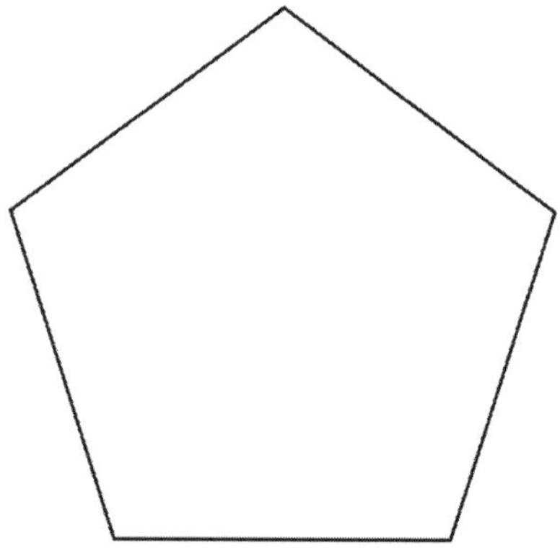

Figure 110. An image showing pentagon.

Source: Image by Wikimedia Commons.

Pentagram

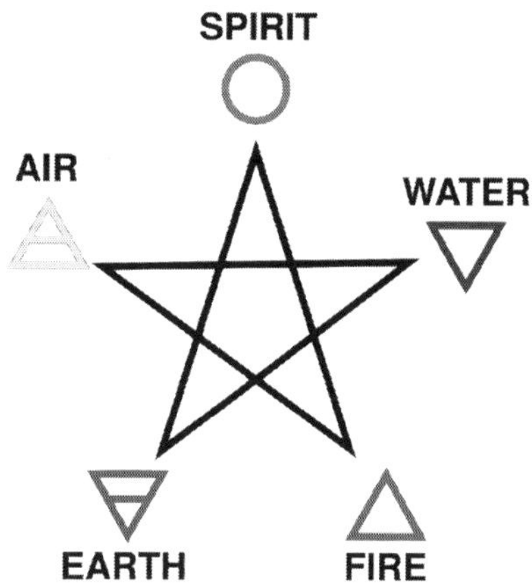

Figure 111. Five elements and pentagram colored.

Source: Image by Wikimedia Commons.

The pentagram is a star-shaped figure made by drawing all of a normal pentagon's diagonals. The pentagram, like the regular pentagon, has five axes of symmetry and rotational symmetry of order five. A pentagram can be formed by extending the sides of a standard pentagon with straight lines.

Pentahedron

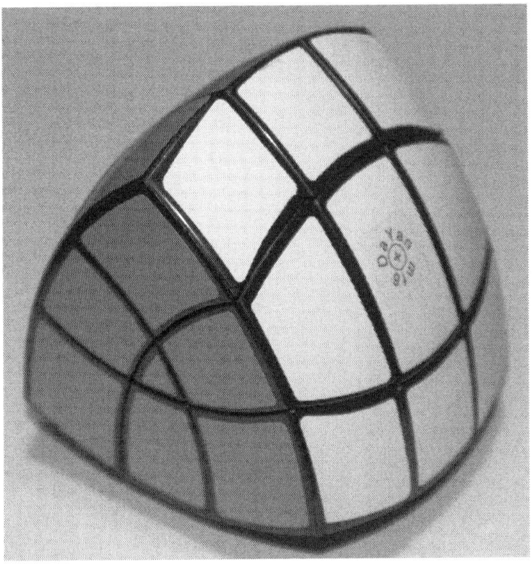

Figure 112. Pentahedron solved cube meister.

Source: Image by Wikimedia Commons.

A pentahedron is a polyhedron with five sides. There is no such thing as a regular pentahedron. In Euclidean 3-space, there are two pentahedrons that can exist. A square pyramid is one of them.

Pentominoes

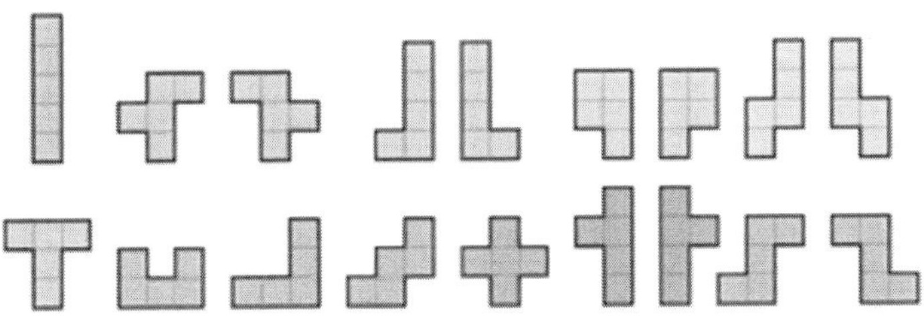

Figure 113. All 18 pentominoes.

Source: Image by Wikimedia Commons.

A pentomino is formed when five squares are linked edge to edge. Each pair of squares must be joined along the entire edge. There are a total of 12 possible ways to put five squares together, each of which is referred to as a pentomino. The nets for an open box are made up of eight pentominoes. Keep in mind that types like the three in figure an are considered to be the same pentomino.

The 12 pentominoes have been sketched, and the squares that make up the open box bases have been shaded.

Percentage

A percentage is a figure or ratio that can be stated as a fraction of 100 in mathematics. If we need to calculate a percentage of a number, we divide it by 100 and multiply it by the whole amount. As a result, the percentage denotes a fraction of a percent. Per cent denotes one hundredth of a percent. The symbol "percent" is used to symbolize it.

There is no dimension to percentages. As a result, it is known as a dimensionless number. If we say 50% of a number, we are referring to 50% of the whole.

Percentages can also be expressed as decimals or fractions, for example, 0.6%, 0.25%, and so on. Academically, the grades earned in any subject are expressed as a percentage. Ram, for example, received 78% on his final exam. As a result, this percentage is derived by adding Ram's total marks in all subjects to his total marks.

Percentage Error

When we measure something in an analysis process, we use percent mistakes to show how big our errors are. We're getting near to the accepted or original figure when the percent mistakes are low. A 1% mistake, for example, implies that we were extremely close to the acceptable number, whereas a 48% error indicates that we were very far from the genuine value. Because hands can shake, materials can be imprecise, or our tools just don't have the potential to estimate precisely, measurement errors are often unavoidable. The percent error formula will tell us how much these unavoidable errors influenced our final results.

Percent Error Formula

The formula for percent error is:

$$PE = (|\text{Estimated value} - \text{Actual value}|/\text{Actual value}) \times 100$$

Or

Here,

T = True or Actual value

E = Estimated value

Percentiles

Percentiles divide a collection of data into 100 equal parts and are frequently used to show how well a person performed in comparison to others in a competition or an examination. A percentile is one of the 100 components into which the data is divided. Assume Luke is taking an exam, and there are 200 other students taking the same exam. Their findings are shown in the figure as a cumulative frequency graph. Let's say we want to know what the 90th percentile examination mark is. The class consists of 200 pupils, and 90% of 200 students equals 180.

It means that 90% of the students, or 180 pupils, did not achieve a score of at least 75% on the exam. Furthermore, it signifies that 10% of the pupils received a grade higher than 75%.

Perfect Number

A perfect number is a whole number that is equal to the sum of all of its factors except itself. The first perfect number, for example, is 6. 6 has the following factors: 1, 2, 3, and 6: 6 = 1 + 2 + 3. Because 6 is the sum of all of its parts except itself, it is a perfect number.

Perfect Square

A number that can be represented as the product of two equal integers is called a perfect square. Because it is the product of two equal integers, 5 5 = 25, 25 is a perfect square. However, because it cannot be written as the product of two equal numbers, 21 is not a perfect square. (7 + 3) = 21

Perfect squares are numbers that can be made by squaring any whole number.

Perimeter

The perimeter of a closed curve or polygon can refer to the closed curve's or polygon's boundary line, or it can refer to the overall length of the boundary line. The total length of the boundary line will be discussed in this entry.

The perimeter of the rectangle in figure a, which is created on a 1-cm grid, is calculated by adding the lengths of the four sides, which equals the total length of the boundary line: $4 + 3 + 4 + 3 = 14$ Perimeter

The rectangle's perimeter is 14 cm.

An alternative method for finding the perimeter of the rectangle is to add together 4 and 3 cm and double the answer. For curved shapes like the circle, it is difficult to find the perimeter by measuring the length of the boundary line, so we use a formula for the perimeter. The circumference of a circle is sometimes referred to as its perimeter.

Period

In mathematics, the length of a periodic function from one peak to the next (or from any point to the next matching point). To put it another way, the length of a whole cycle.

Periodic

A body is considered to be in periodic motion if the motion it is doing is repeated at equal time intervals, such as a rocking chair or a swing in motion. A periodic function is defined as follows:

A function that returns to the same value on a regular basis.

Despite the fact that periodic motion and oscillatory motion sound the same, not all periodic movements are oscillatory motion. The primary distinction between a periodic motion and an oscillatory motion is that a periodic motion is applicable to any motion that repeats over time, but an oscillatory motion is only applicable to motions that execute around an equilibrium point or between two states. All periodic motions can be defined by a periodic function.

Permutations

When the order of the arrangements counts, a permutation is a mathematical technique that specifies the number of alternative arrangements in a collection. A common mathematical problem involves selecting only a few items from a group of items in a specific sequence.

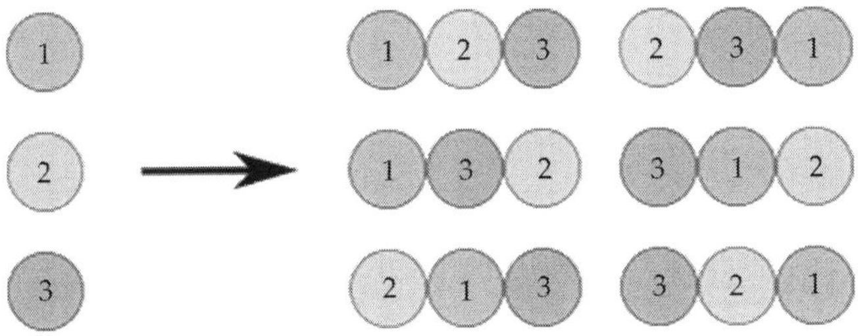

Figure 114. Permutations.

Source: Image by Wikimedia Commons.

Perpendicular Bisector

A perpendicular bisector is a line segment that meets another line perpendicularly and divides it into two equal halves. When two lines connect in such a way that they make a 90-degree angle with one other, they are said to be perpendicular to each other. A bisector, on the other hand, cuts a line into two equal halves. As a result, a perpendicular bisector of a line segment AB denotes that it intersects AB at 90 degrees and divides it into two equal halves.

Perpendicular Bisector Properties

- It separates or bisects AB into two equal parts.
- It is perpendicular to (or forms a right angle with) AB.
- Every point in the perpendicular bisector is equidistant between points A and B.

When working with practical geometry, you will frequently encounter the use of perpendicular bisectors, such as when attempting to construct an isosceles triangle or determining the center of a circle. The procedures for making a perpendicular bisector of a line with a compass and a ruler are shown below.

How Do You Make a Perpendicular Bisector?

You'll need a ruler and compasses. The stages for constructing a perpendicular bisector of a line segment are as follows:

Step 1: Create a PQ line segment.

Step 2: Set the compass to a length somewhat greater than half the length of PQ.

Step 3: With the compass pointer positioned at point P, draw arcs above and below the line.

Step 4: Keeping the compass length constant, position the compass pointer at point Q. Draw two arcs above and below the line while maintaining the compass pointer at Q.

Step 5: Label the spots where the opposing arcs intersect as X and Y.

Step 6: Using a ruler, draw a line connecting X and Y.

The perpendicular bisector bisects PQ at point J, implying that the length PJ equals JQ. The angle formed by the two lines is 90 degrees.

Perpendicular Lines

- Are lines that are perpendicular to each other.
- Two straight lines are perpendicular if they are at right angles to each other, i.e., the angle between them is 90 degrees. For example, if two lines are perpendicular and one is horizontal, the other is vertical.

PI

The sign for pi is, which is a Greek alphabet letter. The value of can be found in a calculator as = 3.141592653… but cannot be written down exactly. is an irrational number, which means it can never be written as a fraction or expressed as a recurring decimal. The fraction 22/7 is sometimes used as a value for, however when represented as a decimal, it is only accurate to two decimal points. Certain lines or lyrics might be used to assist you recall how important it is.

The following is an explanation for the origin of: Assume you have a circular object, say a dinner plate, and you want to measure its circumference (C) and diameter (D) with a length of thread and a ruler.

The definition of π is $\pi = C/D$.

Pictogram

A pictogram is a type of statistical graph that is used to depict discontinuous data. The graph is made up of a sequence of pictures, each of which represents a different amount of data.

Pie Graph

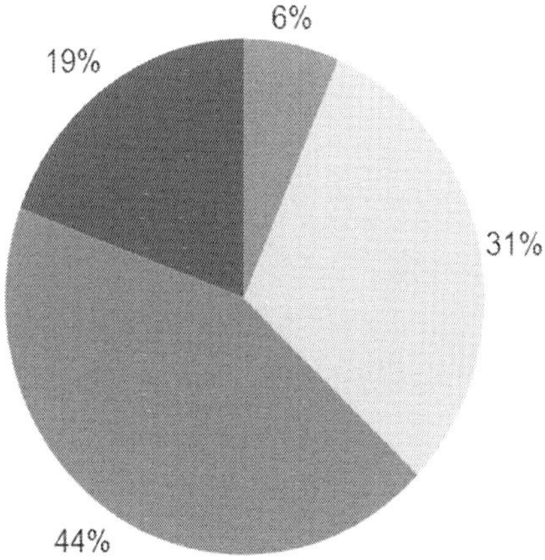

Figure 115. An image showing pie chart.

Source: Image by Wikimedia Commons.

A pie graph, often known as a pie chart, is a type of statistical graph used to depict discrete data. A pie graph, unlike a bar graph, lacks axes. It is a circular graphic with sections within the circle. In the circle graph, each piece of data is represented by a sector of the circle, whereas in the bar graph, each piece of data is represented by a column. A data item's frequency is proportional to the angle in the sector that represents it. t. The frequency is also proportional to the sector's area. The height of the column in the bar graph is related to the frequency. When creating a pie graph, we take advantage of the fact that there are 360 degrees in a full rotation.

Pint

A pint is a unit of measurement for volume, or the amount of liquid that a container can store. A pint is equal to half a quart. A quart is another unit of measurement that equals one-quarter gallon. Consider money to figure out how much a quarter of a gallon is. A gallon is equivalent to a dollar, and a quart is equivalent to a quarter. A gallon has four quarts because there are four quarters in a dollar.

Placeholder

A placeholder is a zero added into a number to keep the numerical value when we round.

Plane

- A plane extends in two dimensions in a geometry.
- It doesn't have any width.
- An example of a co-ordinate geometry plane can be seen.
- The coordinates define the points position in the aircraft.

In Mathematics, a plane is an endlessly long, two-dimensional surface.

A plane is an analogous to two dimensions which can be composed of a point, a line and an area of three dimensions.

Flat bands can look like subspaces of a few more spaces, as the walls of the room extend extraordinarily far.

As under the context of Euclidean geometry, these walls experience a separate existence by themselves.

What's an issue?

One point is in a plane that does not have a dimension, i.e., no width, length or depth.

What's a track?

A line is a series of endless points in opposite directions. It simply has one measurement, i.e., length. The points on the same line are known as hill points.

Plane in Algebra

The points in the algebra are shown in the coordinate plane, which is an example of a geometric plane. The coordinate plane has a numerical line that extends continuously from left to right and another that extends infinitely up and down. The entire co-ordinate plane cannot be viewed.

Arrows indicate that the x-axis and y-axis on the end of numerical lines extend continuously. These numbers are twin lines, where the plane stretches forever. If we draw the graph in a plane, then there is no thickness of the dot or line.

Platonic Solids

Plato, who lived around 400 BC, was the inspiration for the Platonic solids. They are also known as regular polyhedra or solids. A Platonic solid has every face be the same regular polygon, and so all of its edges and vertices are the same. There are only five platonic solids, and Plato defined them and demonstrated how to build the solids from their nets. The five platonic solids are as follows:

- Cube composed of six congruent squares.
- Dodecahedron composed of 12 regular congruent pentagons.
- Icosahedron composed of 20 congruent equilateral triangles.
- Eight congruent equilateral triangles form a regular octahedron.
- Four congruent equilateral triangles form a regular tetrahedron.

Crystals form in the form of some regular polyhedra. Crystals of ordinary salt, for example, develop like cubes, while crystals of chrome alum grow as regular octahedrons.

Point of Inflection

A function is a specific relationship between two sets in mathematics (input set and output set). Every member of the output set is related to one or more members of the input set in a unique way. The function is denoted by the letter "f." There are various types of functions. They are classified based on the categories. The nature of the graph is one such category. The functions can be classified into two groups based on the nature of the graph.

- The Convex Function
- Function of Concavity

Concavity and convexity can both occur once or multiple times in a function. The point at which the function is neither concave nor convex is referred to as the inflection point or point of inflection.

The point of inflection, also known as the inflection point, is the point at which the function's concavity changes. It denotes a change in function from concave down to concave up or vice versa. In other terms, an inflection point is the point at which the slope's rate of change changes from increasing to decreasing or vice versa. Those are most emphatically not local maxima or minima. They are fixed points.

Polygon

A polygon is a closed plane shape with straight sides that do not intersect. The sides meet at places known as vertices. The polygon is said to be regular when all of its sides are the same length and all of its angles are the same size. The sum of the interior angles of a polygon has a formula, which is deduced as follows.

The triangle has a total of 180 pounds of three interior angles. Even 180 krono, a + b + c. Two triangles are a quadrilateral with the shared vertex V: 2 plywood 180 linen Equals 360 cloth (which is the angle sum of a quadrilateral). A common vertex V comprises the three triangles of a Pentagon: 3 — 180 live = 540 livres, the Pentagon angle sum. By pattern of the triangle, quadrilateral and pentagon, we can extend the procedure to encompass the n-side polygon. A polygon on the n-sided composes (n−2) triangles with a common vertex. The sum of a polygon with n faces has a n − 2 = 180 pounds. The wording is a sum of the inner corners of a polygon with n-sides = (n − 2).

Polyhedron

The plural of polyhedron is polyhedra. A polyhedron is a strong figure with polygon planes. The faces meet at the ends, while the rims meet at the tops.

Polynomial

Polynomials consist of algebraic terms consisting of coefficients and variables. Also, sometimes indeterminates are called variables. For polynomial expressions, but not by division, we can conduct arithmetic operations, such addition, suppression, multiplication as well as positive integer exponents. An example is x2+x–12, which includes a polynomial. There are three words: x2, x and –12 in this example. The word polynomial is derived from the Greek words 'poly' meaning 'many' and 'nominal' means 'terms.' A polynomial may have several terms, but not endless. Polynomial consists of two terms: Poly (meaning "many") and Nominal, which are referred to as "terms," respectively. An expression consisting of variables, constants and exponents combined using mathematical operations such as addition, subtraction, multiplication and partition is defined as a polynomial (No division operation by a variable). It is categorized as monomial, binomial and trinomial, based on the number of terms present in the phrase.

The following are examples of consistencies, variables and exponents:

Constants. E.g., 1, 2, 3, etc.

Variables. E.g., g, H, X, Y and so forth.

Exponents: For instance: 5 in x^5 etc.

The function polynomial is referred to as P(x), where x is the variable. $P(x) = x^2–5x+11$, for instance

The function will be p if a variable is indicated (a).

Degree of a polynomial

The polynomial degree is defined as the monomial's highest degree in a polynomial. Therefore, a polynomial equation with the biggest exponent is called a polynomial degree.

Polyominoes

The name of a polyonminus is given to figures produced by squares at the margins of the plane. Polyominoes are often described as groups with a number of characteristic sides that ignore rotations and reflection. For example, the monomino is a single square trival group. Almost anyone knows the form of the domino, a rectangle made up of two squares. Slightly more fascinating are the trominoes: this group contains two members, the "right tromino" and the "right (angle) tromino."

Four-square polyominoes, and five of those squares, are called tetrominoes. The iconic Tetris arcade puzzle game requires players to tie these polyominos together while leaving as few trousers as possible.

The pentominoes are the five-square polyominoes. There are 12 different pentominoes. For simplicity, everyone can be considered like an alphabet letter and so a "letter name" is provided.

Population

Population is a term used in statistics and refers to the whole data set when a sample is chosen. You must be clear about the population of your sample before choosing a sample. For example, if you chose a sample of kids to complete a questionnaire on how you feel about having an agent on the Board of Trustees, all of the students in your school will presumably be inhabitants. Your population would presumably be all the students in the country if you worked for the examining board and chose a collection of students to gain some general information on the exam results. Not all the pupils in the world as a population would be a good concept, because all of them would have not understood the same thing.

Positive Integers

Positive entries actually belong to a wider number group known as integers.

All the numbers, positive and negative, are integer. We mean numbers without fractions or decimals with entire numbers. You can name your 'counting numbers' also positive integer, as they are identical. You don't have fractions or decimals or negatives.

In one line, all numbers on the right of the null are positive values. Positive entries are numbers in the world that you observe around you. There is positive integer signage for speed limits on all our roadways. The amount of dishes needed is a positive integer when you set the table for dinner. The same applies to the quantity of family and guest chairs required. In math, the numbers you see are positive, not fractions or decimal figures. It's the simple numbers.

1 346 8 78 7 485 98 7 225 2 6 11 1 346 8 78 7 485 34 98 7 225 2 6

All the above numbers are positive. Note that they have neither a negative symbol in front nor fractions nor decimals. These are the numbers with which you can work, because the laws of fractions or decimals do not need to be worried.

Pound

The pound is a weight unit in the imperial unit system. The shortcut is lb.

Powers

The power of a number indicates how often to multiply the number. Subjects or indexes are sometimes known as power. For instance, 8^2, "8 at power 2" or "8 at power 2" or just "8 squared," might be named.

Some interesting truths regarding power:

- If the indicators are 1, then you have only the number. 5^1 = 5, for instance
- If the indexes are 0, you will receive 1. 5^0 = 1 for instance
- Exponents make writing and using numerous multiplications easier

- Negative exponent means how often to divide the number one by one.

Prime Factor

A factor that's a prime number.

In other words, one of the prime numbers, to give the original number, can be multiplied.

Example: 15 are primarily 3 or 5 (for 3 is primary 5, 3 is primary). (and 3 is primary 5).

Prime Number

A prime number is one that can only be divided by itself and one without leaving any remainders.

A prime number is a number bigger than one that has only two elements — itself and one.

A prime number cannot be divided by another number without a remainder.

13 is an example of a prime number. It can only be split by one and thirteen. When you divide a prime number by another number, you get numbers left over, for example, 13 6 = 2 remainder 1.

Principal

The term "principal" has numerous financial connotations. The most widely used term refers to the initial amount of money borrowed in a loan or invested. It can also refer to the face value of a bond, as with the former. The term "principal" can also refer to a particular person or parties, the owner of a private corporation, or the primary participant in a transaction.

Prism

A prism is a polyhedron (solid) with congruent and parallel bases and tops, as well as parallel and identical cross sections to the base. The prisms we shall look at are upright prisms, also known as "right" prisms. Some prisms have names based on the geometry of their cross section. The cross section can be a regular polygon, such as the cube, or a nonregular prism, such as the trapezium prism. Take note that the trapezium prism is shaped like a swimming pool with a deep end and a shallow end.

A Prism's Volume the length of a prism is the distance between its cross-sectional ends. The volume is given by Volume of prism = area of cross section × length

Probability of An Event

Probability is a form of ratio in which we evaluate the number of times an event can occur to the total number of possible outcomes.

Probability equals the number of desired outcomes divided by the number of available possibilities.

Another way to answer the question, "What are the odds or likelihood that an event will occur?" is to consider the probability of an event occurring. If we roll a die, it has an equal probability of falling on any of the six numbers (1, 2, 3, 4, 5, or 6), as long as it is not biased. It will undoubtedly land on one of these numbers, but which one?

The chances of it landing on the number three are one in every six. In principle, if you rolled the die six times, it should land once on a 3 and once on each of the other five numbers. In practice, though, the dice may not land on a 3 at all or it may land on a 3 every time in six rolls! All we can say is that if we roll a die, the probability of it falling on a 3 is 1/6, assuming that all of the numbers on the die are equally likely to appear.

Probability Tree

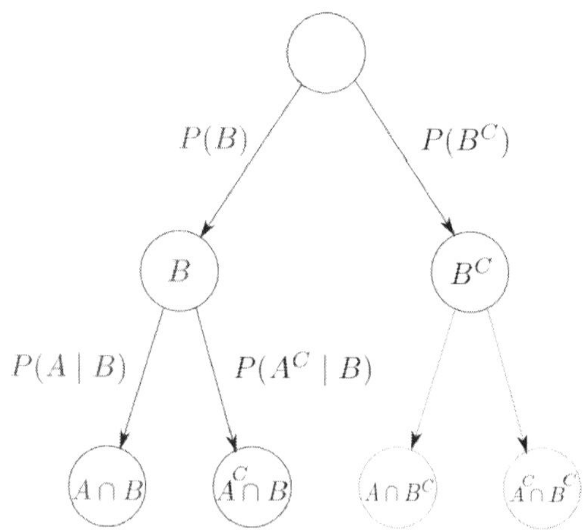

Figure 116. Probability tree.

Source: Image by Wikimedia Commons.

Probability trees can be used to compute combined probability for event sequences. It allows you to graphically map out the probabilities of many different alternatives without the need for sophisticated probability formulas.

Product

The outcome of one or more multiplications is referred to as a "product." The mathematical equation ab=c, for example, might be understood as "a times b equals c," where c represents the product.

More broadly, the product of many various types of mathematical objects, including some that are not numbers, is possible. The Cartesian product, for example, gives the product of two sets. The product topology in topology can be used to define the product of spaces. The direct product returns the product of two groups, vector spaces, or modules. The category product is used in category theory to compute the product of objects.

Proof

To demonstrate the truth of a proposition, give a logical argument that establishes the assertion as a fact. A geometry theorem is proven using a logical argument that is frequently based on axioms or fundamental geometry theorems.

Proportion

Ratio and direct proportion are two words that are related yet frequently confused. Before proceeding, you may want to review the entry Ratio. When two ratios are equivalent, they are said to be in direct proportion. Having a photograph magnified is a good approach to explain straight proportion.

enlarged.

We say two figures are similar if they are in direct proportion. This suggests that two figures are comparable if one of them is a magnified version of the other. Aside from similar figures, other quantities can be in direct proportion. Hooke's law states that the extension of a spring is proportional to the weight that produces it. Inverse proportion is a type of proportion that we will now contrast with direct proportion. Extending a spring by adding weights to it is an example of direct proportion. When one variable (the weight) grows, the other quantity (the spring extension) increases proportionally.

Protractor

A protractor is a flat, transparent plastic semicircle with two scales for measuring angles in degrees on it. Each scale ranges from 0 to 180, and each marker on the scale equals one. Figure a shows the 10 markings but not the 1 markings. The protractor will measure angles to a precision of one degree. The straight edge of the semicircle is known as the center, and the center of the semicircle is known as the straight edge. A protractor can be totally circular as well.

Each angle is made up of two arms, or rays, that meet at a point, or vertex. The vertex of the angle is O in figure b, while the arms are OA and OB. When measuring an angle with a protractor, we set the protractor's center on the angle's vertex O and the straight edge on one of the angle's arms, as illustrated in the image. We set the protractor's straight edge on the OA arm (or we could use the arm OB).

Starting at the angle's arm where the straight edge is set, we check the two scales on the protractor and see that the inside scale begins at zero and the outer scale begins at 180. We measure around the scale using the inside scale, which starts at zero, until we reach the opposite arm OB. At 123, the OB arm crosses the inner scale. The angle AOB is 123 in size.

Always use the scale that starts at zero on the arm of the angle on which the straight edge is placed. If the angle's arms are too short to reach the scale on the protractor, they must be stretched before utilizing the protractor.

Pyramid

This is a solid object (a polyhedron) with a base (any polygon) and edges traced from each vertex of the base that meet at a single point V known as the pyramid's apex. A pyramid's base can be any polygon, but its other faces are invariably triangles. We will look at the correct pyramids as well as those with regular polygon bases. A pyramid is distinguished by the shape of its base. The pyramid depicted in figure a has a square base.

Pythagoras Theorem

This theorem is named after the 500-year-old Greek mathematician Pythagoras. The theorem, which applies to any right-angled triangle, is a formula that connects the areas of squares that can be drawn on the triangle's sides. The shaded triangle in figure an is right-angled, and squares A, B, and

C are drawn on the triangle's sides. Area of square C Equals area of square A Plus area of square B, according to Pythagoras' theorem.

"The square on the hypotenuse of a right-angled triangle is equal to the sum of the squares of the other two sides," we declare. Pythagoras' theorem is used to find the length of one side of a right-angled triangle if the lengths of the other two sides are known.

Pythagoras Triples

Pythagorean triples are defined as a2+b2 = c2, where a, b, and c are three positive numbers. These triples are denoted as (a,b,c). The perpendicular is a, the base is b, and the hypotenuse is c of a right-angled triangle. The most well-known and tiniest triplets are (3,4,5). For further information, read Pythagoras' theorem

Mathematician Pythagoras was interested in mathematics, physics, and philosophy. He was born in 570 BC in Greece. He is best known for the Pythagorean Theorem, which states that triangles with a right angle have 900 angles. The hypotenuse is the side opposite the right angle of a right-angled triangle. Side p is the shorter of the two sides adjacent to the right angle.

Q

Quadrants

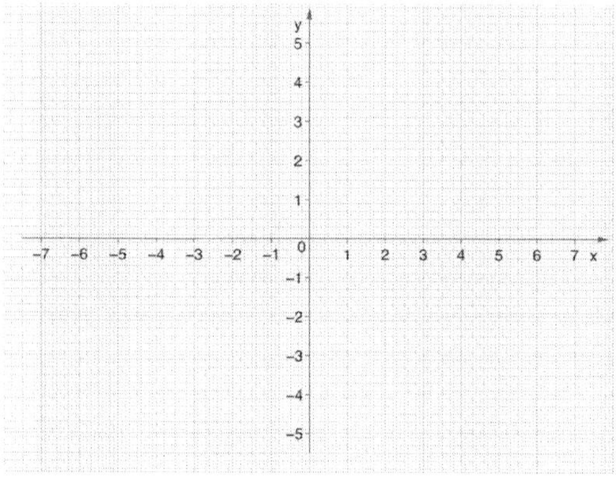

Figure 117. Coordinate system four quadrant.

Source: Image by Pixabay.

The axis of a two-dimensional Cartesian plane system divides the plane into four infinite regions is called quadrants. These regions encompass both positive and negative values of x and y. The crossroads of two lines are known as the point of reference. This is the reference point (or starting point) in the coordinate system for all measurement. the graph is divided into sections or four quadrants, based on those values.

1st Quadrant: The upper right-hand corner of the graph is the first quadrant. In this quadrant the values of x and y both are positive.

2nd Quadrant: The upper left-hand corner of the graph is the second quadrant. In this quadrant, the value of x is negative whereas the value of y is positive.

3rd Quadrant: The lower left-hand corner of the graph is the third quadrant. It contains the negative values of both x and y.

4th Quadrant: Finally, the fourth quarter is at the lower right-hand corner, that has a positive value of x and negative values of y.

Equations of QUADRATIC quality

The polynomial equations of degree 2 in a single type of variable f(x)=ax2+bx+c where a, b, c, = R and a -kin 0. Quadratic equations It is the general form of a four-way equation in which an is called the leading coefficient and c is referred to as the absolute term f. (x). The quadratic equation x values are the roots (α,β) of the quadratic equation.

There are always two roots in the quadratic equation. The roots can be actual or imagined in nature.

A quadratic polynomial becomes a quadratic equation if it is equal to zero. The values of x which satisfy the equation are known as the quadratic equation roots.

$Ax^2 + bx + c = 0$ in general

E.g., $3x^2 + x + 5 = 0$, $+ 7x+5 = 0$, $x^2 + x = 0. = 5$

Quadratic Formula

Often, "$ax^2 + bx + c= 0$" can easily be resolved by factoring x, setting each component equal to zero, and then solving each factor. But sometimes the quadratic is too chaotic, or factoring is not at all, or you don't want to factor. The Quadratic Formula can always find the solution; however, it may not always be successful.

The Quadratic Formula employs the letters "a," "b," and "c" from "$ax^2 + bx + c$," where "a," "b," and "c" are simply numbers; they are the "numerical coefficients" of the quadratic equation you've been given to solve.

The Quadratic Formula is developed from the square-completion procedure and is explicitly stated as:

The Quadratic Equation:

For $ax^2 + bx + c = 0$, the values of x that constitute the equation's solutions are provided by:

Quadratic Graph

A parabola is the graph of the quadratic expression y = axe 2 + bx + c, where b or c, or both, may be zero. In this formula, an equals zero, and a, b, and c are all real values. The parabolas we analyses can be "cup"-shaped, with a greater than zero, or "hat"-shaped, with a less than zero.

Quadrilateral

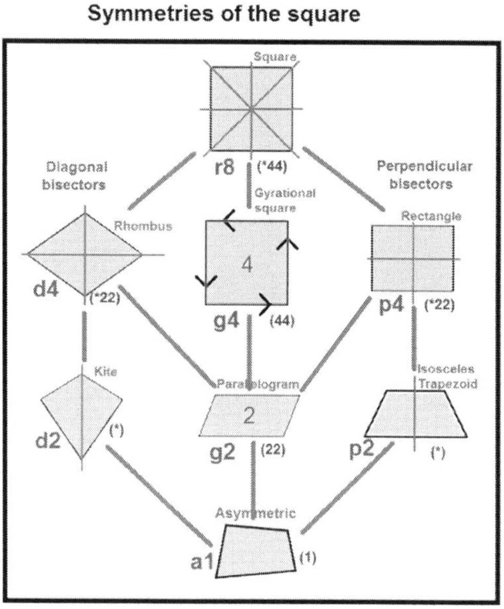

Figure 118. Different types of quadrilateral.

Source: Image by Wikimedia Commons.

The number of sides of a form or item is used in geometry to classify it. Shapes are classified into three types: triangles, rectangles, and squares (3-sides)

The quadrilateral (4-sides)

The Pentagon (5-sides)

Hexagonal (6-sides)

Hexagon (7-sides)

Octagons (8-sides) and so forth.

A quadrilateral is a plane shape with four sides or edges and four corners or vertices. Quadrilaterals can be normal shapes having four sides, such as rectangles, squares, trapezoids, and kites, or they can be irregular and uncharacterized.

Another approach to categorize quadrilaterals is as follows:

Convex Quadrilaterals: A quadrilateral's diagonals are totally contained within a figure.

Concave Quadrilaterals: At least one of the diagonals is partially or completely outside of the figure.

Intersecting Quadrilaterals: Intersecting quadrilaterals are not just quadrilaterals with non-adjacent sides that intersect. These quadrilaterals are referred to as self-intersecting or crossed quadrilaterals.

Quadruple

This term translates as "four times." For example, assume John invested $100 and doubled his investment in five years. This indicates that his investment has grown fourfold to the value of 4, $100 = $400.

Qualitative Data

Data that can only be expressed in words rather than figures, such as the colors of automobiles in a parking lot.

Qualitative Graphics

The data used in most graphs are quantitative data regarding numbers and their relationships. Descriptive data, such as eye color, gender, nationality, attitudes, and so on, are examples of qualitative data. When we examine qualitative graphs, we are looking for correlations between descriptive data components, notably attitudes.

Quantitative Data

Quantitative data is described as data in the form of counts or numbers, with each data-set having a unique numerical value. This data is any quantifiable information that can be utilized for mathematical computations and statistical analysis, allowing for real-world decisions to be made based on these mathematical derivations. Quantitative data is used to answer queries like "How many?" "How often?" and "How much?" This information may be validated and easily evaluated using mathematical procedures.

There are quantities that correspond to various factors, for example, "How much did that laptop cost?" is a question that will collect quantitative data. Most measuring parameters have values connected with them, such as pounds or kilograms for weight, dollars for cost, and so on.

Because of the ease of mathematical derivation that quantitative data provides, it makes measuring various parameters controllable. Quantitative data is typically collected for statistical analysis through the use of surveys, polls, or questionnaires distributed to a specified segment of a population. The obtained results can be applied to a population.

Questionnaire

A survey is an investigation into anything, such as how people feel about the sale of alcohol in supermarkets. A survey could take the form of a questionnaire that a sample of the population fills out. The sample information is utilized to generate predictions and draw conclusions about the entire population. For example, based on the questionnaire replies, it may be feasible to deduce that selling alcohol in supermarkets is not popular with the general public. The questionnaire will be designed to collect information that accurately represents the public's opinion about the sale of alcohol, rather than to develop a biased view in favor of wine shop owners! The questionnaire is a collection of written questions, not a verbal interview.

Quotient

The quotient is the result of dividing one integer by another. For example, if we divide 6 by 3, the resulting number is 2, which is the quotient. It is the outcome of the division process. The quotient can be either an integer or a decimal. For exact divisions such as $10 / 5 = 2$, the quotient is an integer, however for divisions such as $12 / 5 = 2.4$, the quotient is a decimal. The remainder of the division is the decimal part of the quotient in the division process with decimal quotient as the response.

In general, knowing the quotient can assist you grasp the size of the dividend in relation to the divisor. Dividend by divisor is the quotient formula. A quotient can be greater than the dividend but less than the divisor.

Division is a process of repeated subtraction. The quotient is equal to the number of times subtraction is performed. The division is represented by a mathematical symbol() consisting of a short horizontal line with a dot above and below the line. The quotient is the result of this division procedure.

R

Radian

A unit of plane angular measurement equivalent to the angle at the center of a circle subtended by an arc whose length equals the radius or approximately 57.3 degrees. In simple terms, is a unit of measure for angles determined by the radius of a circle, for example, while taking the length of the radius and wrapping it around a circle, the angle formed thereafter at the center of the circle by this arc is equal to 1 radian.

Radius

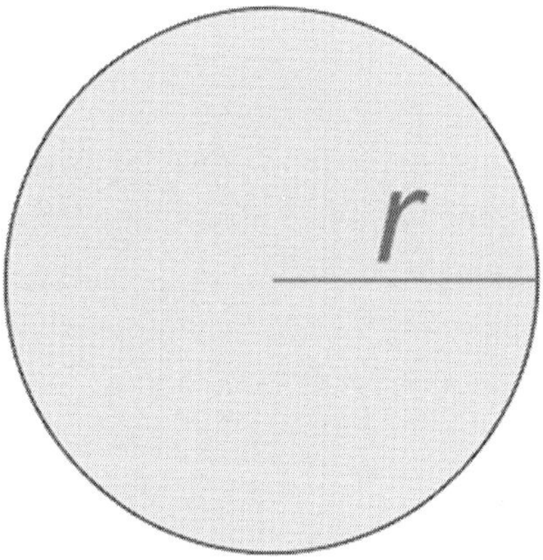

Figure 119. An image showing radius of a circle.

Source: Image by Wikimedia Commons.

Circles are by far the most recurrent shapes found in everyday life. It is widely regarded as one of the most critical topics in classical geometry. Radius is by far the most critical part of a circle. It is the distance between the centers of the circle to any point on its boundary. It is a line segment extending from the center of a circle or sphere to the circumference or bounding surface. In other words, when the center of a given circle is connected to any point on its boundary or circumference by drawing a straight line, that line is considered as the radius of that circle.

There can be more than one radius as there are innumerable points to establish measurement on the circumference of a circle. A circle can have more than one radius because there are infinite points on its circumference; known as radii. Therefore, a circle has an infinite number of radii and all the radii are at an equal distance from the center of the circle.

The formula for the radius of a circle, when the diameter is known is: Radius = Diameter/2. In cases, where the circumference is known, the formula for the radius is: Radius = Circumference/2π. In cases where the area is known, the formula for the radius is: Radius = √(Area of the circle/π).

Random Numbers

In mathematics and statistics, random numbers are numbers chosen as if by chance from some specified distribution such that selection of a large set of these numbers reproduces the underlying distribution. Random numbers are numbers that occur without any design or pattern. They are numbers that occur completely by chance. For example, in the case of a dice, when it is rolled, all numbers that appear on it, namely, 1, 2, 3, 4, 5, and 6, all of these have an equal chance of showing up, unless the dice is not tampered with. Therefore, the dice will only display random numbers from 1 to 6.

Therefore, a random number is a number picked out from a set of limited or unlimited numbers that has not fixed pattern for prediction. The pool of numbers is almost always independent from each other. However, the pool of numbers may follow a specific distribution. For example, the height of the students in a school tends to follow a normal distribution around the median height. If the height of a student is picked at random, the picked number has higher chance to be around the median height instead of being categorized as either very tall or very short.

Random Sample

It is a statistical term, used for defining a subset of statistical population in which each member of the subset is accorded an equal probability of being chosen. Therefore, a random sample is part of a population and is chosen free of bias from the whole population, and is expected to have the same characteristics as the whole population. In simple terms, a random sample is basically an unbiased representation of a larger group or population.

A simple example of a simple random sample would be picking out the names of 25 employees out of a hat from a company of 250 employees. In such a case, the population is 250, being represented by all the employees, since every employee has an equal probability of being picked out. In science, random sampling is employed to conduct randomized control tests or in case of blinded experiments.

Range

In mathematics, the difference between the highest and the lowest values is defined as the range. It is a statistical measurement of dispersion, or simply the amount of a given data set when stretched out from smallest to largest.

Range is one of four simple tools of statistics:

- **Mean** – The arithmetic means or the average, is simply the sum total of a collection of numbers divided by the count of numbers in the collection.
- **Median** – The statistical median is the value separating the higher half of a data sample from the lower half.
- **Mode** – The mode is the most frequently occurring value in a data set.
- **Range** – The range of a data set is the mathematical difference between the largest and smallest value.

The first three statistical tools are measures of central tendency, or how similar the numbers are while range highlights a measure of dispersion; it shows how different the numbers are.

Rate

In mathematics, a rate signifies the ratio that compares two different quantities with different units. For instance, in case of a car traveling at a speed of 120 miles in 3 hours; it can be expressed as 120 miles divided by 3 hours which means 40 miles per hour. Thus, 40 miles/hour is an example of a unit rate. A few common examples of rate are: time rate or average speed (miles/hour), cost price i.e., cost/dollar and other general rates such literacy rate, birth rate and so on.

Rate of Change

The rate of change (ROC) is the speed at which a variable changes over a specific period of time. It is the rate that depicts how one quantity changes or alters in relation to another quantity. The ROC is often illustrated by the Greek letter delta.

In mathematics, the rate of change is often employed to explain the percentage change in value over a defined period of time, and it represents the momentum of a variable. The calculation for ROC is simple in that it takes the current value of a stock or index and divides it by the value from an earlier period. Subtract one and multiply the resulting number by 100 to give it a percentage representation, i.e.,

ROC= (previous value/current value−1) * 100

Rate of change is a critical financial theory as it indicates the security momentum and other such trends to potential investors. For instance, a security with high momentum, or one that has a positive ROC, generally surpasses the market in the short term. On the other hand, a security with an ROC that falls below its moving average or one that has a low or negative ROC is more likely to have a diminished value and can hint as being a sell signal to investors.

Rate of change is also an excellent yardstick to gauge the market bubbles. Even in cases where the momentum is good and traders scout for securities with a positive ROC, if a broad-market ETF, index, or mutual fund experiences a sharp increase in its ROC in the short term, it may be a sign that the market is unsustainable. If the ROC of an index or other broad-market security is over 50%, investors should be wary of a bubble.

Ratio

In mathematics, a ratio is a comparison of two or more numbers that indicates their sizes in relation to each other. A ratio compares two quantities by division, with the dividend or number being divided termed the antecedent and the divisor or number that is dividing termed the consequent. In simpler terms, it is a term that is used when we have to compare two or more than two numbers.

For example, if in basket there are eight melons and six lemons, the ratio of melons to lemons is eight to six i.e., 8:6, which is equivalent to the ratio 4:3.

Rational Expression

Rational expression is an expression that is the ratio of two polynomials. It implies that both the numerator and denominator are polynomials in it. It is similar to a fraction as it is also a ratio of algebraic expression, which consists of an unknown variable.

A rational function is the ratio of two polynomials P(x) and Q(x) like this

f (x)= P (x)/Q(x)

Except that Q(x) cannot be zero (and anywhere that Q(x) =0 is undefined).

In order to find out the root or zero of polynomial expression, they must be considered equal to zero. But in order to find the zeros of rational function or expressions only the numerator equal to zero has to be put when the expression has been reduced to its lowest terms. By lowest terms it implies that there are no common factors between the numerator and denominator. Like in the case of a fraction, say 3/9, it is not in the lowest form. It can be further reduced by taking 3 as a common factor. Thus, 1/3 is the lowest form. Likewise, for example, $x^2+2x/3x$ is a rational expression, which is not in its lowest form. Yet, if we consider the common factor x from both numerator and denominator, we get (x+2)/3, which is the lowest form of the expression.

Rational Numbers

In mathematics, a rational number is a type of real number, that is of the form of p/q where p and q are integers and q is not equal to zero. In simpler terms, if a number is expressed as a fraction where both the numerator and the denominator are integers, the number is a rational number. The following are a few examples of rational numbers:

- ½
- –3/5
- 0.5 or 5/10
- –0.9 or –9/10
- 0.131313 or 13/99

Real Numbers

In mathematics, real numbers are the numbers that include both the rational and irrational numbers, in the number system. Real numbers can be defined as the union of both the rational and irrational numbers. They include both the positive and negative integers and fractions and irrational numbers and are represented by the symbol "R." All the natural numbers, decimals and fractions fall under real numbers.

Reciprocal

The reciprocal of any given number when multiplied by that number gives one as a product. Thus, it is also called the multiplicative inverse. In mathematics, reciprocal can be defined as the inverse of a number or a value. For a real number n, reciprocal is 1/n, such as reciprocal of 3 is 1/3. Therefore, here we convert the number into an upside-down form. For example, the reciprocal of 7 is 1 divided by 7, i.e., 1/7. Now, if we multiply a number by its reciprocal, it gives a value equal to 1. Thus, it is also called multiplicative inverse.

The word reciprocal comes from the Latin word "reciprocus" which means "returning." Hence, it returns its original value, if we take the reciprocal of an inverted number. In Mathematics, reciprocal means an expression which when multiplied by another expression, gives unity (1) as a result. The reciprocal of any quantity is, one divided by that quantity. For any number 'a,' the reciprocal will be 1/a.

Example: Reciprocal of a number 8 is 1/8. Thus, a reciprocal is basically a flipped number that can bring back the original number. With the exception of 0, every number has a reciprocal since there is nothing that can be multiplied by 0 to create a product of 1, thus, it has no reciprocal.

Rectangle

Figure 120. Basic rectangle.

Source: Image by Wikimedia Commons.

In geometry, a rectangle is simply a four-sided polygon, known as a quadrilateral, which has its opposite sides parallel and four right angles. A general term for rectangle is oblong. A rectangle is a two-dimensional shape that has equal opposite sides. It also has two axes of symmetry and rotational symmetry of order two.

Every rectangle forms two diagonals that intersect in the middle are of equal length. Both pairs of "opposite" triangles formed by the two diagonals are congruent isosceles triangles. When all sides of a rectangle are of the same length, it forms a special type of rectangle called a 'square.' The word "rectangle" comes from the Latin word "rect," which means "right," and the Latin word "angulus," which means "angle." Each corner of a rectangle has a right angle of 90°.

Reflection

In geometry, reflection is known as a flip. It is simply the geometrical transformation that occurs when a shape, called the object, is reflected in a mirror to obtain the image. Thus, a reflection is simply, the mirror image of a shape. For example, the mirror image of the small Latin letter p for a reflection with respect to a vertical axis would look like q. Its image by reflection in a horizontal axis would look like b. A reflection maps every point of a figure to an image across a fixed line. The fixed line is called the line of reflection. A reflection is an involution: when applied twice in succession, every point returns to its original location, and every geometrical object is restored to its original state. Reflection is the only transformation that changes sense.

Properties of reflection

The mirror line is an invariant line. Any point on the mirror line is an invariant point. The perpendicular distance of a point from the mirror line is equal to the perpendicular distance of its image from the mirror line. The straight line that joins a point to its image is at right angles to the mirror line. Angle sizes are invariant. Area is invariant. The object and image are congruent shapes. Reflection is an indirect transformation.

Reflex Angle

A reflex angle refers to all angles measured above 180 degrees (half circle) and below 360 degrees (full circle). A reflex angle will always have either an obtuse or an acute angle on the other side of it. For instance, 195 degrees, 220 degrees, 330 degrees, etc. are all reflex angles, whereas 180° and 360° are not reflex angles as they are both straight angles and complete angles respectively.

Regions

Regions and closed regions are often used as domains of functions or differential equations.

According to Kreyszig, "A region is a set consisting of a domain plus, perhaps, some or all of its boundary points. (The reader is warned that some authors use the term "region" for what we call a domain [following standard terminology], and others make no distinction between the two terms)."

According to Yue Kuen Kwok, "An open connected set is called an open region or domain. ...to an open region we may add none, some, or all its limit points, and simply call the new set a region."

Relation

A relation between two sets is a collection of ordered pairs containing one object from each set. If the object x is from the first set and the object y is from the second set, then the objects are said to be related if the ordered pair (x, y) is in the relation.

A function is a type of relation. But, a relation is allowed to have the object x in the first set to be related to more than one object in the second set. Therefore, a relation may not be represented by a function machine, because, given the object x to the input of the machine, the machine couldn't spit out a unique output object that is paired to x. There are eight main types of relations which include: empty relation, universal relation, identity relation, inverse relation, reflexive relation, symmetric relation, transitive relation, and equivalence relation.

Remainder

In mathematics, remainder refers to the amount left over after the division process. In arithmetic, the remainder is the integer "left over" after dividing one integer by another to produce an integer quotient (integer division). For example, if 7 pens were to be divided among 4 children, 3 pens would be left over and when translated into mathematical terms, the remaining 3 pens are the remainder.

In another example, if 20 are to be divided is to be divided by number 7, the quotient is 2, and the remainder is 6. As a rule, the remainder is always lesser than the divisor. During many calculations, the remainders are ignored or rounded off, to give only a whole number answer. In the decimal number

7.02, the number 2 after the decimal is the remainder and is sometimes ignored to give only the whole number answer 5.

Revolution

A revolution is a complete turn, which is an angle of 360° or four right angles.

Rhombus

A rhombus can be best defined as a special parallelogram and a quadrilateral, with four equal parallel sides. Though a rhombus has four equal sides, it is also a type of parallelogram as it has four congruent sides. Though all rhombi or rhombuses are parallelograms, but all parallelograms are not rhombuses. A rhombus is also known as a diamond, lozenge, or a rhomb.

Some of the important properties of a rhombus are as follows:

- A rhombus has equal sides.
- The opposite sides of a rhombus are parallel.
- Opposite angles of a rhombus are equal.
- In a rhombus, diagonals bisect each other at right angles.
- Diagonals bisect the angles of a rhombus.
- The sum of two adjacent angles is equal to 180 degrees.
- The two diagonals of a rhombus form four right-angled triangles which are congruent to each other.
- You will get a rectangle when you join the midpoint of the sides.
- You will get another rhombus when you join the midpoints of half the diagonal.
- Around a rhombus, there can be no circumscribing circle.
- Within a rhombus, there can be no inscribing circle.

Right Angle

In the field of geometry and trigonometry, a right angle is an angle forming an accurate 90°, corresponding to a quarter turn. A right angle can be in any orientation or rotation as long as the internal angle is 90°. It is by far, the most frequently viewed angle in daily life, and can be found everywhere from the corners of a room, to the screen of a tablet or mobile phone, the edges of boxes and so on. Rectangles and squares both form right angles on their sides.

A right angle is formed when two rays meet at a vertex exactly equal to 90 degrees. It implies that both the rays are perpendicular to each other.

Right Prism

It is a prism in which the sides are at right angles to the uniform cross section. The right triangular prism has the uniform cross section which is at right angles to each of the three sides; it has rectangular sides. A cylinder is another example of a right prism.

Right Pyramid

A right pyramid is a pyramid with a base that is a regular polygon and whose apex is directly above the center of the base. Any regular pyramid is called a right pyramid if the line between the apex and the center of the base is perpendicular to the base. Cones are an example of right pyramids.

Right Triangle

A right triangle or a right-angled triangle is a triangle; that has one of its interior angles equal to 90 degrees or any one angle is a right angle. Therefore, this triangle is also known as the right triangle or 90-degree triangle. It is an important concept of trigonometry.

In a right-angled triangle, one of its angles is that of 90 degrees while the other two angles sum up to 90 degrees. The sides that include the right angle are perpendicular and the base of the triangle. The third side is called the hypotenuse, which is the longest side of all three sides. The side opposite to the right angle is the smallest side.

The three sides of the right triangle are related to each other. This relationship is explained by Pythagoras theorem. According to this theorem, in a right triangle,

Hypotenuse2 = Perpendicular2 + Base2

Thus, a right triangle is a three-sided closed shape and has one perpendicular side.

Roman Numerals

Roman numerals are any of the symbols used in a system of numerical notation based on the ancient Roman system. The symbols are I, V, X, L, C, D, and M, standing respectively for 1, 5, 10, 50, 100, 500, and 1,000 in the Hindu-Arabic numeral system. A symbol placed after another of equal or greater value adds its value. A symbol placed before one of greater value subtracts its value. A bar placed over a number multiplies its value by 1,000.

There are 4 basic principles for writing Roman numerals as listed below:

- The letters I, X, C can be repeated thrice in succession. Additionally, L, V, D cannot be repeated or the number is considered to be invalid.
- In case a lower value digit is written to the left of a higher value digit, it is subtracted.
- In case a lower value digit is written to the right of a higher value digit, it is added.
- Only I, X, and C can be used as subtractive numerals.

Roots

The root of a number x is another number, which when multiplied by itself a given number of times, equals x. For example, the second root of 9 is 3, because 3x3 = 9. The second root is usually called the square root. When we solve an equation we obtain solutions, and these solutions are sometimes called the roots of the equation. For example, $x = 4$ is a root of the linear equation $2x - 3 = 5$, and $x = 3$ and $x = -4$ are the roots of the quadratic equation $x^2 + x - 12 = 0$.

Rotation

A geometrical concept, a rotation refers to a transformation that turns a figure about a fixed point, called the center of rotation. A rotation turns a shape around. Any rotation is a motion of a certain space that preserves at least one point. While an object and its rotation are the same shape and size, but the figures may be turned in different directions. Rotations can be either clockwise or counterclockwise. An example is a child's carousel at a fair. The carousel spins about its center O, which is fixed. The carousel can turn clockwise or counterclockwise through a full turn or through an angle which may be part of a full turn.

S

Sample

In statistics and quantitative research methodology, a sample refers to a set of individuals or objects selected from a statistical population.in simple terms, it refers to a part of the population. It is simply the specific group used to collect data. Therefore, the size of the sample is always lesser than the size of the total population. In case of random samples, it is chosen without any partiality from the whole population, and, it is expected to share similar characteristics with the whole population.

Sample Space

A concept of probability theory, a sample space, also known as sample description space or possibility space, is a collection or a set of possible outcomes of a random experiment. The symbol used to denote sample space is "S." The subset of possible outcomes of an experiment is called events. A sample space may contain a number of outcomes which depends on the experiment. In cases wherein it contains a finite number of outcomes, it is known as discrete or finite sample spaces.

The samples spaces for a random experiment are written within curly braces "{ }."

For example, Sample Space for a single coin: Constructing a sample space for the experiment that consists of tossing a single coin: the outcomes could be labeled h for heads and t for tails. Then the sample space is the set: S= {h, t}.

Scale

In mathematics, the term scale is used to depict the relationship between a measurement on a model and the corresponding measurement on the actual object. In simple terms, The ratio of the length in a drawing (or model) to the length on the real thing.

Scales are used in our everyday lives, using them for measuring lengths with a ruler or measuring tape, temperature using a thermometer, angles in degrees using a protractor, weight we use scales using weighing scales, revolutions per minute, using a revolution counter volts, using a voltmeter, altitude using an altimeter and so on.

Scale Drawing

A drawing that shows a real object with accurate sizes reduced or enlarged by a certain amount (called the scale).

Scalene Triangle

Figure 121. Scalene triangle.

Source: Image by Flickr.com.

A scalene triangle is a triangle in which all three sides are of varying lengths and the angles of a scalene triangle have different measures. Irrespective of the differing measurements, the sum of all the interior angles of the scalene triangle is not impacted and always adds up to 180 degrees, which satisfies the angle sum property of the triangle.

Properties of scalene triangle

The following are the properties of a scalene triangle:

- It has three differently measuring sides.
- It has three angles all of which are unequal.
- It has no parallel or equal sides and hence no line of symmetry.
- It has no line or point of symmetry.
- The angle opposite to the longest side would be the greatest angle and vice versa.
- The interior angles of the triangle can be acute, obtuse, or right angles. Thus, a scalene triangle can be an obtuse-angled, acute-angled, or right-angled triangle.
- In the case of an acute scalene triangle, the center of the circumscribing circle will lie inside the triangle.

- In an obtuse scalene triangle, the circumcenter will lie outside the triangle.

Scatter Diagram

A scatter diagram, also called scatter plot, scatter graph, X-Y graph and correlation chart is a statistical tool used to analyze if two variables are related to each other. The scatter diagram graphs pairs of numerical data, with one variable on each axis, to look for a relationship between them. If the variables are correlated, the points will fall along a line or curve. The better the correlation, the tighter the points will hug the line. The two variables are the two coordinates of a point, and the points are plotted on a Cartesian graph.

A scatter diagram is used when:

- There is paired numerical data,
- When the dependent variable may have multiple values for each value of your independent variable,
- When trying to determine whether the two variables are related, such as: When trying to identify potential root causes of problems; After brainstorming causes and effects using a fishbone diagram to determine objectively whether a particular cause and effect are related; When determining whether two effects that appear to be related both occur with the same cause; When testing for autocorrelation before constructing a control chart.

Scientific Notation

Scientific notation is a method of expressing numbers that are either too large or too small to be written in a decimal form. It is a unique way of writing numbers and makes big and small values easy to denote. Scientific notation helps scientists handle complex and very large numbers or very small numbers. For example, instead of writing 0.0000000076, we write 7.6 x 10–9. In scientific notation, the digit term indicates the number of significant figures in the number. The exponential term only places the decimal point. For example, the scientific notation of 46,600,000 is 4.66 x 10^7.

The scientific notation helps us to represent the numbers which are very huge or very tiny in a form of multiplication of single-digit numbers and 10 raised to the power of the respective exponent. The exponent is positive if the number is very large and it is negative if the number is very small.

Score

This is a set of 20. It is the natural number that succeeds 19 and precedes 21. For example, a score of soldiers is a set of 20 soldiers.

Second

A second is a very small unit of angle measure. In the measurement of angles, a second is 1/60th of a minute, and a minute is 1/60th of a degree.

Sector of A Circle

The sector of a circle is the amount of space enclosed within the boundary of a sector. A sector of a circle always originates from the center of the circle. A sector of a circle is defined as the portion of a circle that is enclosed between its two radii and the arc adjoining them. The semi-circle is the most common sector of a circle, which represents half of a circle

The sector is a portion of a circle that can be defined based on the three points mentioned below:

- A portion of a circle covered by two radii and an arc.
- A circle is divided into two sectors and the divided parts are known as minor sectors and major sectors.
- The large portion of the circle is the major sector whereas the smaller portion is the minor sector.
- In the case of semi-circles, the circle is divided into two equal-size sectors.

The 2 radii meet at the part of the circumference of a circle known as an arc, formed a sector of a circle.

Segment of A Circle

A segment of a circle is the region that is bounded by a chord of the circle and intercepted by an arc of the circle. A segment with an intercepted arc less than a semicircle is called a minor segment. A sector with an intercepted arc greater than a semi-circle is called a major segment. When something is divided into parts, each part is referred to as a segment. In the same way, a segment is a part of the circle.

Properties of Segment of Circle

The properties of a segment of a circle are:

- It is the area that is enclosed by a chord and an arc.
- The angle subtended by the segment at the center of the circle is same as the angle subtended by the corresponding arc. This angle is usually known as the central angle.
- A minor segment is obtained by removing the corresponding major segment from the total area of the circle.
- A major segment is obtained by removing the corresponding minor segment from the total area of the circle.
- A semicircle is the largest segment in any circle formed by the diameter and the corresponding arc.

Area of a Segment of Circle

An arc and two radii of a circle form a sector. These two radii and the chord of the segment together form a triangle. Thus, the area of a segment of a circle can be calculated by subtracting the area of the triangle from the area of the sector. i.e.,

Area of a segment of circle = area of the sector – area of the triangle

Semicircle

A semicircle is a half – circle, formed by cutting a whole circle along a diameter line. Any diameter of a circle cuts it into two equal semicircles. The formula for the area of a semicircle is A = pi * r^2/2, where A stands for area and r stands for the circle's radius.

Septagon

In geometry, a septagon or heptagon is a polygon with seven sides and seven angles.

The properties of a regular septagon are as follows:

- Its seven sides, seven verticals and seven interior angles.
- It has 14 diagonals.
- The sum of all its interior angles is 900 degrees.
- The sum of its exterior angles is 360 degrees.
- Each interior angle of a regular septagon is 128.571 degrees.

SI Units

On the foundation of seven defining constant the SI rest on and the cesium hyperfine splitting frequency, the speed of light in vacuum, the Planck constant, the elementary charge (i.e., the charge on a proton), the Boltzmann constant, the Avogadro constant, and the luminous efficacy of a specified monochromatic source. By using an explicit-constant formulation and experimentally realized using a specific mises en pratique (practical technique) SI base units are expressed.

The seven SI base units, which are comprised of:

- Length – meter (m)
- Time – second (s)
- Amount of substance – mole (mole)
- Electric current – ampere (A)
- Temperature – kelvin (K)
- Luminous intensity – candela (cd)
- Mass – kilogram (kg)

For the measurement purpose the International System of Units (SI), commonly known as the metric system and it is the international standard for measurement. On May 20, 1875 the International Treaty of the Meter was signed in Paris and by seventeen countries, including the United States and is now celebrated around the globe as World Metrology Day.

There are various international bodies which are being established by the Meter Convention: CGPM – General Conference on Weights and Measures; CIPM – International Committee for Weights and Measures; and BIPM – The International Bureau of Weights and Measures, as it is on the NIST and also provides official U.S. representation.

It has been said that with the special names and symbols the SI is made up of 7 base units that define the 22 derived units. In the international commerce the SI plays an essential role and which is commonly used in scientific and technological research and development.

Side

It can be defined as one of the line segment which heal in forming a polygon. For instance, one side of the triangle ABC is the line segment AB. Alternatively, a side can refer to one of the faces of a polyhedron. For example, the shaded region of the cuboid is called a side of the cuboid. The cuboid has six sides.

Sieve of Eratosthenes

It has been seen that up to any given limit Sieve of Eratosthenes is a simple and ancient algorithm used to find the prime numbers.

Similar Figures

The two figures are said to be similar at the time when one plane figure is an enlargement of another plane figure. It has been said that if two figures are similar, they are the same shape; for example, both figures may be triangles, or both figures may be hexagons. Similar figures have the following properties:

In the other figure the angles in one figure are equal to the corresponding angles and in the two figures corresponding sides are in the same ratio.

Simultaneous Equations

At the same time simultaneous equations are two or more equations that are to be solved. In two variables simultaneous equations gets deal here and are sets of two equations, which also represent the unknown activities. It has been noted that at the same time solution to the simultaneous equations is a value for each of the two variables that make both of the equations true, therefore the word simultaneous.

Sine

The sine is a trigonometric function of an angle in mathematics. In the context of a right triangle the sine of an acute angle is defined for the specified angle, it is the ratio of the length of the side that is opposite that angle, to the length of the longest side of the triangle.

Sine Rule

With the sines of its angles the sine rule is a set of trigonometric formulas connecting the lengths of the three sides of a triangle. In the triangles it is mainly used that are not right-angled, because simpler methods are used for right-angled triangles.

For the sine rule the notation is being used and also for the cosine rule, is that the sizes of the three angles of the triangle are referred to using capital letters, say A, B, and C, and the lengths of the sides are small letters, a, b, and c. In order to use the sine rule and the cosine rule one need to be familiar with this notation.

Skew Lines

In the different planes skew lines are the one which are not parallel and also don't intersect. Parallel lines are the lines which are in the same plane and don't get intersect. Moreover, in mathematics always remember that lines extend forever in both directions. To visualize them one need to think in 3-D. Therefore, to illustrate three-dimensional concepts on paper or a computer screen it is often difficult.

Slant Height of Cone

Along the curved surface the slant height of an object (such as a cone, or pyramid) is the distance which is drawn from the edge at the top to a point on the circumference of the circle at the base. In general term, it can be defined as the shortest possible distance from the base to the apex along the surface of the solid, denoted either as s or l. In the calculation of the slant height in any object the slant height formula helps.

Solids

It is considered as the three-dimensional object. There are some of the solids which are made of flat surfaces called planes, such as the cube, the cuboid, the pyramid, and the tetrahedron. These are known as polyhedrons, and have faces, vertices, and edges that are related by Euler's formula. It has been said that some of the solids are made up of a mixture of flat surfaces and curved surfaces, such as the hemisphere, the cone, and the cylinder and other solids are made up of only curved surfaces, such as the sphere, the torus, and the ellipsoid.

Sphere

It has been noted that from its center this is a solid shape in which every point on its surface is the same distance. The same distance is the radius of the sphere, A ball can be used as the perfect example of the sphere. The earth is approximately a sphere. Any cross section of a sphere is a circle.

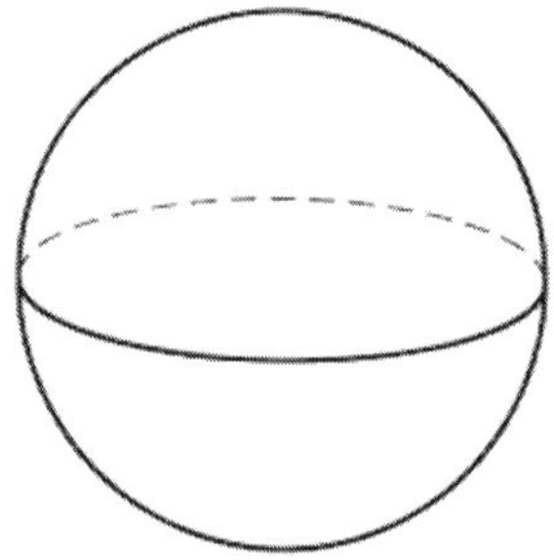

Figure 122. Sphere.

Source: Image by Wikimedia Commons.

Spiral

By a point that moves around a fixed point and at an ever-increasing distance from the fixed point at that a spiral is a curve traced out. It has been said that it is the effect produced when a thick piece of rubber or carpet is rolled up. Spirals occur in nature; for example, some seashells are in the shape of spirals.

Square

With each of its four angles equal to 90° and all four sides equal in length a square is a quadrilateral. IT has been seen that the square has four axes of symmetry and has a rotational symmetry of order four. At the right angles the diagonals of the square bisect each other, forming four congruent right-angled isosceles triangles. With the sides of the square the diagonals make angles of 45°.

Square Numbers

This is the set of numbers that are the squares of the natural numbers. To square a number, everyone should multiply the number by itself. For example, $5^2 = 5 \times 5 = 25$.

Square Root

In the mathematics this refereed as an operation that occurs frequently and especially in the use of the theorem of Pythagoras. In this the square root operation is being undo at the time when one takes the square root of a number (or a term). It has been said that taking the square root is the inverse operation of squaring. For example, $5^2 = 25$, and the square root of 25 is 5. The square root of 25 is written as $\sqrt{25}$. The square root of a number can also be written in index form: for example, $\sqrt{25}$ can be written as $25^{1/2}$. Using a scientific calculator square roots of numbers can be found.

Standard Deviation

A standard deviation is defined as the statistic that measures the dispersion of a dataset relative to its mean. By determining each data point's deviation relative to the mean the standard deviation is being calculated as the square root of variance. There is a higher deviation within the data set if the data points are further from the mean, thus, the more spread out the data, the higher the standard deviation.

In finance standard deviation is referred as the statistical measurement, when applied to the annual rate of return of an investment, sheds light on that investment's historical volatility. The greater the standard deviation of securities, the greater the variance between each price and the mean, which shows a larger price range. For example, a volatile stock has a high standard deviation, while the deviation of a stable blue-chip stock is usually rather low.

Standard Form

By the name scientific notation standard form is sometimes called. In the shortened form it is a convenient way of writing very large or very small numbers. In science, this especially useful, which often deals with very large and small numbers. In astronomy there is a unit of distance called the "light year," which is approximately equal to 6,000,000,000,000 miles. It cannot easily be compared with other numbers of similar size when it is written in that form it takes up a lot of room and cannot. In standard form a number is written as a × 10n, where n is an integer and a is a number between 1 and 10, or, more precisely $1 \leq a < 10$. This makes the size of a clear: it can equal 1, but cannot equal 10.

Star Polygons

By joining the vertices of regular polygons these are star shapes formed. Regular star polygons were well known to Pythagoras and his followers, who lived about 500 years BC. In fact, the pentagram was adopted as the badge of the ancient Pythagorean school.

Pentagon

With a straight line to the third vertex the first vertex is joined, the second vertex is joined to the fourth vertex, the third vertex to the fifth vertex, and so on. Then the polygon is removed, and the result is a star pentagon, which is also known as a pentagram.

Hexagon

For the hexagon the process is being repeated.

Octagon

To third the vertex the first octagon the first vertex is drawn and the second to the fourth, the third to the fifth, and so on. In the second octagon the first vertex is joined to the fourth vertex, the second to the fifth, the third to the sixth, and so on.

Stationary Point

In terms of the gradient of the curve at that point a stationary point on a curve is defined. To make the clear understanding of the term one must see the entry Gradient of a Curve. A stationary point on a curve is the point where the gradient of the tangent to the curve is zero.

Statistics

In order to draw conclusions or make predictions this is a branch of mathematics that is about the collection, display, and analysis of numerical data.

Stem And Leaf Graph

This is a graph which is being used in the statistics and is really a histogram in which each item of the data is shown in the graph. The stem and leaf graph has the appearance of a histogram, but none of the original data are lost.

Substitution

By a number substitution usually means replacing a variable in a formula or equation that represents the variable, or by another variable. In the entries given in the references are being described as the various examples of the substitutions. One type is explained here, which is replacement by numbers of the variables in algebraic terms.

Subtend

In circle geometry theorems subtend is a word which is used. on an arc of a circle a subtended angle is an angle that stands. The arc AB referred to is highlighted in the figure. The angle x may be subtended at a point on the circumference of the circle or at its center.

Surface

A well-known solid is a cube, which has six square faces, and these six faces make up the surface of the cube and a surface is the complete boundary of a solid.

Surface Area

It has been defined as the surface area of a solid is the total area of its surface.

T

Tangent and Radius Theorem

Suppose that T is a point on a circle with center O, and a tangent is drawn to the circle at that point T. If a radius of the circle is also drawn through the same point T, then the angle between the tangent and the radius is a right angle. One could say that the radius is perpendicular to the tangent.

Tangents From a Common Point

From one point to a circle this is a geometry theorem about two tangents drawn. The point common to the two tangents PT and PQ is the point P. The theorem states that the two tangents are equal in length. In figure a, PT = PQ, and this theorem is stated as follows:

- Tangents from a common point are equal.

Tangram

It is known as the Chinese puzzle. Into the seven shapes a square is being cut. The object of the puzzle is to cut out the seven shapes and rearrange all of them to form interesting figures of animals, humans, etc.

Temperature

A measure of how hot or cold something temperature is. The two scales of temperature in common use are Fahrenheit (F) and Celsius (C). Celsius used to be called centigrade. The units for measuring temperature are degrees, but degrees Celsius (°C) are not the same size as degrees Fahrenheit (°F). Conversions between the two temperatures can be done using formulas, as follows. To change °C to °F we use the formula $F = 9/5C + 32$.

Tessellations

As the mosaic's tessellations are known. A flat surface is being covered by a tessellation which is a tiling pattern. The tiles we often use are regular polygons such as equilateral triangles, squares, or regular hexagons, but a variety of shapes may be used. Some tessellations involve more than one shape, such as combining square tiles with regular octagons. A tessellation is only correct if the tiles fit together with no gaps and no overlapping and a definite repeated pattern can be easily recognized.

Tetrahedron

With the four plane faces each of which is a triangle is a pyramid. The prefix "tetra" means four. One of the four faces is called the base, and the vertex that is opposite to the base is called the apex. The volume of a tetrahedron = 1/3 area of base × altitude.

Theorem

By the steps of reasoning a theorem is a statement or the formula that can be proved. By Euclid begins the study of geometry as set out with a few definitions and axioms (statements that we believe are obviously true and do not require proof) and uses them to prove theorems, which in turn are used to prove more theorems, and so on, building a chain. For an example on proving a geometry theorem refer to the entry proof.

Time Series

As the time changes, in this entry a time series is taken to be a statistical graph that shows how a quantity varies. For example, during the first 6 months of a given year a time series may be a line graph that shows the changes in temperature in New York City. Other examples of quantities that may be measured are weight, height, cost, age, popularity, and so on. A time series is a record of what has happened in the past and is used so that a comparison can be made with another time series, or to make a prediction about future changes. In the following way a time series may be used. To promote the sales, one must own a supermarket and decide to employ an advertising agency. For a period of 1 year or more and have historical record of how well you are doing it would be a good idea to first measure sales.

Ton

A ton is a unit of weight and is equal to 1000 kilograms. to a building site would expect several tons To arrive a builder having a truckload of sand delivered. One ton of water has a volume of 1 cubic meter at 4°C.

Transformation Geometry

In this bool the transformation geometry is the study of how sets of points and their positions in space are changed by enlargements, reflections, rotations, and translations. On the specific points or the shapes, one should focus and the fact whole of the space is being transformed. The original set of points or shape is called the object, and the set of points or shape that results from the transformation is called the image. Properties of the original shape that do not change after the transformation are called invariant properties. For example, the invariant properties may be area, shape, angle size, and so on. The transformations reflection, rotation, and translation leave size and shape invariant (this includes lengths of lines, the area of the shape, and angles in the shape). Transformations that do this are called isometries. It is clear that enlargement is not an isometry, because it changes the size of the shape.

Translation

In this book a translation is one of four transformations and defined in terms of a grid, such as the squares on a chessboard. Column vectors are used to represent translations.

Trapezium

With one pair of opposite sides parallel a trapezium is a quadrilateral, which means a four-sided polygon. The angles a and b are called cointerior angles and their sum is 180° , which means that a + b = 180°. The two triangles ACD and BCD are equal in area. The trapezium has rotational symmetry of order one. The trapezium on the right is an isosceles trapezium, which has two sides equal in length, two pairs of congruent angles, and one axis of symmetry.

Travel Graphs

As the distance time graphs these are known as and in which time is the horizontal axis and distance is the vertical axis. Travel graphs usually depict a journey.

Tree Diagram

The branches of a tree are designed to show the events of an experiment and enable the probabilities of each event to be calculated which is being explained. In studying the probability, a tree diagram can be used to list the sample space of an experiment, which is a list of all possible outcomes. To calculate the probabilities, it is also being used. In this entry this latter use of tree diagrams that is described. The tree diagram is also recognized as a probability tree.

Trial And Error

To a problem this is a process that involves guessing or estimating the solution and then improving on the guess to obtain a more accurate answer. This is sometimes called trial and improvement, which seems a better description of the process.

Triangle

With the three sides a triangle is a polygon. Under the respective entries the following triangles are studied: isosceles triangle, equilateral triangle, and scalene triangle. Triangles will always tessellate.

Trigonometry

In many entries it gets appears and trigonometry, like Pythagoras' theorem, has a wide variety of uses. So, in this entry a brief introduction is being given and then the trigonometry will be applied to solving right-angled triangles. It is essential to have a scientific calculator available, set in the degree mode.

In the mathematics trigonometry is one of those divisions that helps in finding the angles and missing sides of a triangle with the help of trigonometric ratios. In the radians or degrees, the angles are either measured. The commonly used trigonometry angles are 0°, 30°, 45°, 60° and 90°.

Trinomial

A trinomial is an algebraic expression that contains three terms. Examples of trinomials are $x + 2a - b$, $3ay - y^4 + 2$, $2x^2 - 3x + 5$.

U

Units

The word unit can be defined as the rightmost position in a number or the one's place in mathematics.

Unity

In the English language the word unity carries many meanings, but it is perhaps best known for its most simple and straightforward definition, which is "the state of being one; oneness." While in the field of mathematics the word carries its own unique meaning and the unique use does not stray too far, at least symbolically, from this definition. In fact, in mathematics, unity is simply a synonym for the number "one" (1), the integer between the integers zero (0) and two (2).

The number one (1) represents a single entity and it is our unit of counting. It is the first non-zero number of our natural numbers, which are those numbers used for counting and ordering, and the first of our positive integers or whole numbers. The number 1 is also the first odd number of the natural numbers.

The number one (1) actually goes by several names, unity being just one of them. The number 1 is also known as unit, identity, and multiplicative identity.

Upper Quartile

The upper quartile is the median of the upper half of a data set. By dividing the data set with the median and then dividing the upper half that remains with the median again this is being located and this median of the upper half being the upper quartile.

The formula for the upper quartile is given as:

$$Q_3 = \frac{3}{4}(n+1)^{th}\ Term$$

For ease of writing, the upper quartile will be noted as Q3, also called the third quartile. This formula does not give the value of Q3, rather the term number that will be Q3. *N* represents the number of elements in the data set. For example, if there are 9 elements in the data set, $n = 9$. To use the formula

$(n + 1)$ will equal 10, and then this is multiplied by 3/4 to obtain 7.5. This means the 7.5th term will be Q3, which will be the average of the 7th and 8th terms.

V

Variance

The variance of a set of data is the square of the standard deviation of the data in statistics: Variance = σ^2

Variation

In this text there are two kinds of the variation which are being discussed. One is direct variation, which is also called direct proportion, and the other is inverse variation, which is also called inverse proportion.

Vector

Both the magnitude and the direction are what the vectors are having. Examples of vectors are force, velocity, and acceleration. Speed is not a vector, because it has magnitude, but not direction. Suppose Ken is using a garden roller on the lawn and he is pulling with a force of 200 newtons. The magnitude of the force is being stated but not the direction. At an angle of 30° to the horizontal one may add that ken is pulling at Ken. To describe a vector quantity both the magnitude and the direction are needed. A line segment represents a vector, and the length of the line segment represents the magnitude or size of the vector and the direction of the line segment is the direction of the vector.

Velocity

As the time changes velocity is defined to be the rate at which the displacement of an object is changing. Another common unit is kilometers per hour. If an object is traveling at a constant speed of v meters/second for t seconds and covers d meters, then the formulas connecting these quantities are:

distance = speed × time, speed = distance/time, time = distance/speed

Velocity as time changes. The basic unit of velocity, as of speed, is meters per second.

Vertically Opposite Angles

At a vertex when the two lines intersect then those are the two pairs of the congruent angles. Vertical opposite angles are equal in size and the angles at the vertex that are opposite each other. In this geometry theorem, "vertically" has no reference to the word vertical, but is derived from the word vertex.

Volume

It has been defined as the volume of a solid shape which is a measure of the three-dimensional space it occupies. In the cubic units it is being measured and which written as units 3. One can find the volume of a solid shape, a cuboid, by counting the number of cubes that its three-dimensional space occupies. The volume of a cuboid measuring 3 cm by 2 cm by 3 cm can be found by counting the number of cubic centimeters (abbreviated cm^3) it occupies. There are three layers of cubes, and in each layer, there are six cubes, so the volume of the cuboid is $6 + 6 + 6 = 18$ cm^3.

W

Whole Numbers

Figure 123. An image showing whole numbers.

Source: Image by Public Domain Pictures.

The whole numbers are considered as the part of the number system in which it includes all the positive integers from 0 to infinity. These numbers exist in the number line. Therefore, they are all the real numbers. It has been said that all the whole numbers are real numbers, but not all the real numbers are whole numbers.

The complete set of natural numbers along with '0' are called whole numbers. The examples are: 0, 11, 25, 36, 999, 1200, etc.

Width

It is being defined as the measurement of the distance of a side of an object. Usually this is the shorter side while the length is the longer side.

X

X-Axis

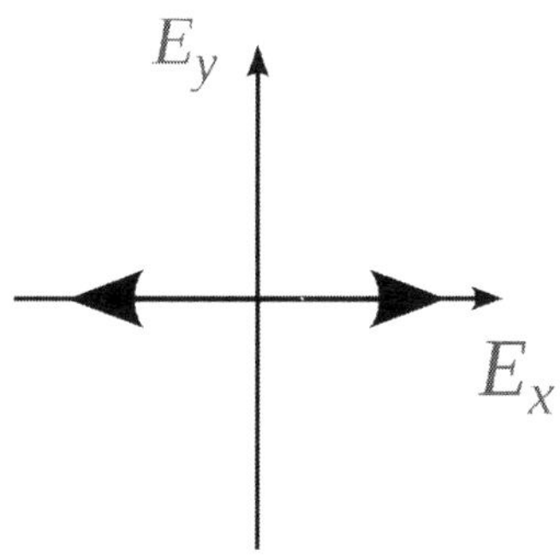

Figure 124. An image showing X axis and Y axis.

Source: Image by Wikimedia Commons.

Through zero the line on the graph that runs horizontally. It is used as a reference line one can measure from it.

X Coordinate

In an ordered pair a ***x*-coordinate** is the *x* value, which is just two mathematical objects, such as numbers, paired together. In that pair usually the x-coordinate the first number. For instance, in the ordered pair (2,5), the *x*-coordinate would be '2.' This value tells you how far a point is from the origin, or starting point, in the *x* direction on a 2-dimensional graph.

Y

Y = mx + b

The general equation of a straight line is y = mx + b, where m is the slope and b is the value where the line cuts the y-axis. This number b is called the intercept on the y-axis.

Yard

In US a unit of length units equal to 3 feet or 36 inches. The abbreviation is: yd. Example: 5 yards can be written 5 yd. One yard is exactly 0.9144 meters In the Metric System (the yard is defined that way).

Z

Zeno's Paradox

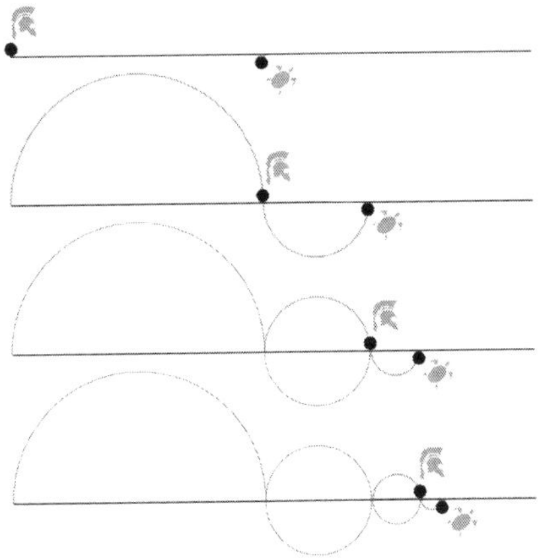

Figure 125. Zeno Achilles Parrado.

Source: Image by Wikimedia Commons.

With the counterintuitive aspects of continuous space and time. Zeno's paradoxes are the set of four paradoxes which are dealing with can converge, so that the infinite number of "half-steps" needed is balanced by the increasingly short amount of time needed to traverse the distances.

BIBLIOGRAPHY

1. Ashino, T., & Yamashita, Y., (2012). Development of an Openmath Content Dictionary for Mathematical Knowledge of Materials Science and Engineering. *Data Science Journal*, [online], *11*, ASMD17–ASMD21. Available at: https://datascience.codata.org/articles/abstract/10.2481/dsj.AMDS–004/ (accessed 16 August 2021).
2. Benoit, A., Chyzak, F., Darrasse, A., Gerhold, S., Mezzarobba, M., & Salvy, B., (2010). *The Dynamic Dictionary of Mathematical Functions* (DDMF). Mathematical Software – ICMS 2010, [online], 35–41. Available at: https://www.researchgate.net/publication/221333704_The_Dynamic_Dictionary_of_Mathematical_Functions_DDMF (accessed 16 August 2021).
3. Bishop, C., (1993). Mathematics dictionary (5th edition) by James and James. Pp 548. (1992). ISBN 0–442–01241–1 (Van Nostrand Reinhold). *The Mathematical Gazette*, [online], *77*(479), 276–277. Available at: https://www.cambridge.org/core/journals/mathematical-gazette/article/abs/mathematics-dictionary–5th-edition-by-james-and-james-pp–548–1995–1992-isbn–0442012411-van-nostrand-reinhold/90AC926A508688A8A6CD53F262634154 (accessed 16 August 2021).
4. Boas Jr., R., (1952). Book Review: Dictionary of Mathematical Sciences. *Bulletin of the American Mathematical Society,* [online], *58*(4), 505–508. Available at: https://www.ams.org/journals/bull/1952–58–04/S0002–9904–1952–09632–4/home.html (accessed 16 August 2021).
5. Clapham, C., & Nicholson, J., (2009). *The Concise Oxford Dictionary of Mathematics.* [online], Available at: https://www.oxfordreference.com/

view/10.1093/acref/9780199235940.001.0001/acref–9780199235940 (accessed 16 August 2021).

6. Crank, J., (1962). Dictionary of Mathematics. *Physics Bulletin*, [online], *13*(11), 306–307. Available at: https://iopscience.iop.org/article/10.1088/0031–9112/13/11/015 (accessed 16 August 2021).
7. Cundy, H., (1967). *Dictionary of Mathematics*. By W. Millington, T. A. p, 260. 21s. (1966). (Cassell). *The Mathematical Gazette*, [online], *51*(376), 175–175. Available at: https://www.cambridge.org/core/journals/mathematical-gazette/article/abs/dictionary-of-mathematics-by-w-and-t-a-millington-pp-x–260–21s–1966-cassell/66A7C318B5E522D90FC5D33B2FD21E6C (accessed 16 August 2021).
8. Dieudonne, J., Iyanaga, S., & Kawada, Y., (1979). *Encyclopedic Dictionary of Mathematics. The American Mathematical Monthly*, [online], *86*(3), p.232. Available at: https://www.jstor.org/stable/2321544?origin=crossref (accessed 16 August 2021).
9. Dodd, A., (1990). *The Penguin Dictionary of Mathematics*, edited by John Daintith and R. D. Nelson. p. 350. (1989). ISBN 0–14–051119–9 (Penguin). *The Mathematical Gazette*, [online], *74*(470), 394–395. Available at: https://www.researchgate.net/publication/324432246_The_Penguin_Dictionary_of_Mathematics (accessed 16 August 2021).
10. Dodd, A., (1993). Beyond numeracy: an uncommon dictionary of mathematics, by John Allen Paulos. Pp 285. (1992). ISBN 0–14–014574–5 (Penguin). *The Mathematical Gazette*, [online], *77*(479), 264–264. Available at: https://www.cambridge.org/core/journals/mathematical-gazette/article/abs/beyond-numeracy-an-uncommon-dictionary-of-mathematics-by-john-allen-paulos-pp–285–699–1992-isbn–0140145745-penguin/D905C1AF0180FB2E48B08ED35FD13D7F (accessed 16 August 2021).
11. Efimov, O., (2018). *Russian-English Dictionary of Mathematics*. [online], Available at: https://www.taylorfrancis.com/books/mono/10.1201/9781351076487/russian-english-dictionary-mathematics-oleg-efimov (accessed 16 August 2021).
12. Farmer, D., (1977). Encyclopaedic Dictionary of Mathematics for Engineers and Applied Scientists. *Journal of the Fisheries Research Board of Canada*, [online], *34*(8), 1253–1254. Available at: https://cdnsciencepub.com/doi/10.1139/f77–186 (accessed 16 August 2021).
13. Goddard, L., (1977). Encyclopedic dictionary of mathematics for

engineers and applied scientists. *Endeavour,* [online], *1*(3–4), p.152. Available at: https://www.sciencedirect.com/science/article/abs/pii/016093277790196X?via%3Dihub (accessed 16 August 2021).

14. Goldrich, A., (1958). A Short Dictionary of Mathematics, by C.H. McDowell. Philosophical Library, New York, (1957). vi + 64 pp. *Canadian Mathematical Bulletin,* [online], *1*(2), 139–139. Available at: https://www.cambridge.org/core/journals/canadian-mathematical-bulletin/article/a-short-dictionary-of-mathematics-by-ch-mcdowell-philosophical-library-new-york–1957-vi–64-pp–275/0F70DFCCF0063C30EB7BC9CF25DD97B3 (accessed 16 August 2021).
15. Goodstein, R., (1962). *Dictionary of Mathematics.* Edited by: C. C. T. Baker. pp. 338. 35s. 1961.(George Newnes Ltd. London). The Mathematical Gazette, [online], *46*(358), 334–334. Available at: https://www.cambridge.org/core/journals/mathematical-gazette/article/abs/dictionary-of-mathematics-by-c-c-t-baker-pp–338–35s–1961george-newnes-ltd-london/8931ADBCDAAEBBB336EE120A95722C20 (accessed 16 August 2021).
16. Hill, J., (1977). Encyclopedic dictionary of mathematics for engineers and applied scientists. *Proceedings of the IEEE*, [online], *65*(7), 1103–1103. Available at: https://ieeexplore.ieee.org/document/1454924 (accessed 16 August 2021).
17. Howarth, R., (2017). *Dictionary of Mathematical Geosciences.* [online], Available at: https://link.springer.com/book/10.1007%2F978–3–319–57315–1 (accessed 16 August 2021).
18. James, G., James, R., & Davis, P., (1959). Mathematics Dictionary. *Physics Today*, [online], *12*(10), 50–52. Available at: https://physicstoday.scitation.org/doi/10.1063/1.3060526 (accessed 16 August 2021).
19. James, R., (1992). *Mathematics Dictionary.* [online], Available at: https://www.researchgate.net/publication/346008176_Mathematics_Dictionary (accessed 16 August 2021).
20. Klein, B., (2021). Dictionaries. *Einführung in Python 3,* [online], 83–97. Available at: https://www.hanser-elibrary.com/doi/10.3139/9783446465565.009 (accessed 16 August 2021).
21. Lord, N., (2001). *The Penguin Dictionary of Mathematics* (2nd edn), edited by David Nelson. pp. 461. (2000). ISBN 0 14 051342 6 (Penguin). The Mathematical Gazette, [online], *85*(503), 348–349. Available at: https://www.cambridge.org/core/journals/mathematical-

gazette/article/abs/the-penguin-dictionary-of-mathematics–2nd-edn-edited-by-david-nelson-pp–461–799–2000-isbn–0–14–051342–6-penguin/CFD8BFC648BF696B53E31D7822246551 (accessed 16 August 2021).

22. Lord, N., (2010). *Penguin Dictionary of Mathematics* (4th edn), David Nelson (ed.). pp. 480. (2008). ISBN 978 0141030234 (Penguin). The Mathematical Gazette, [online], *94*(530), 355–355. Available at: https://www.cambridge.org/core/journals/mathematical-gazette/article/abs/penguin-dictionary-of-mathematics–4th-edn-david-nelson-ed-pp–480–1099–2008-isbn–978–0141030234-penguin/DF4F57543CEA33CA3CACAA0F9D7CC474 (accessed 16 August 2021).
23. Maliketi Chiphambo, S., (2019). Mathematics Dictionary: Enhancing Students' Geometrical Vocabulary and Terminology. *Metacognition in Learning,* [online], Available at: https://www.intechopen.com/books/metacognition-in-learning/mathematics-dictionary-enhancing-students-geometrical-vocabulary-and-terminology (accessed 16 August 2021).
24. Mcginty, R., Mutch, W., & Beynen, J., (1985). A Brief Historical Dictionary of Mathematical Terms. *The Mathematics Teacher,* [online], *78*(8), 638–640. Available at: https://pubs.nctm.org/view/journals/mt/78/8/article-p638.xml (accessed 16 August 2021).
25. Nicholson, J., (2014). *The Concise Oxford Dictionary of Mathematics.* [online], Available at: https://www.oxfordreference.com/view/10.1093/acref/9780199679591.001.0001/acref–9780199679591 (accessed 16 August 2021).
26. Parker, C., (1980). A mathematical dictionary, by R. E. Jason Abdelnoor. Pp 125. Flexicover, (1979). SBN 0 08 021862 8/021882 2 (Wheaton). *The Mathematical Gazette*, [online], *64*(428), 130–130. Available at: https://www.cambridge.org/core/journals/mathematical-gazette/article/abs/a-mathematical-dictionary-by-r-e-jason-abdelnoor-pp–125–150-flexicover–275-hard-cover–1979-sbn–0–08–021862–8021882–2-wheaton/94EC3D7932223FEE700646B58FC6D9CF (accessed 16 August 2021).
27. Pearce, J., (1969). Mathematics Dictionary (3rd. Edition). By Glenn James and Robert C. James. pp. viii. 517. 163s. 6d. (1968). (Van Nostrand). *The Mathematical Gazette,* [online], *53*(385), 332–333. Available at: https://www.cambridge.org/core/journals/mathematical-

gazette/article/abs/mathematics-dictionary–3rd-edition-by-glenn-james-and-robert-c-james-pp-viii–517–163s–6d–1968-van-nostrand/C6BE7C3948282C168845115F1F85F739 (accessed 16 August 2021).

28. Pribitkin, E., & James, (1970). Mathematics Dictionary. *The American Mathematical Monthly,* [online], *77*(2), p.206. Available at: https://www.jstor.org/stable/2317359?origin=crossref (accessed 16 August 2021).
29. R. B. H., (1953). *Dictionary of Mathematical Sciences*. I. German-English. By L. Herland. pp. 235. 24s. (1951). (Hafner Publishing Co.). *The Mathematical Gazette*, [online], *37*(320), 136–137. Available at: https://www.researchgate.net/publication/274182656_Dictionary_of_Mathematical_Sciences_I_German-English (accessed 16 August 2021).
30. Ricardo, H., & Schwartzman, S., (1995). The Words of Mathematics: An Etymological Dictionary of Mathematical Terms Used in English. *The American Mathematical Monthly,* [online], *102*(6), p.563. Available at: https://www.jstor.org/stable/2974781?origin=crossref (accessed 16 August 2021).
31. Ricardo, H., (1995). The Words of Mathematics: An Etymological Dictionary of Mathematical Terms Used in English. By Steven Schwartzman. *The American Mathematical Monthly*, [online], *102*(6), 563–565. Available at: https://www.tandfonline.com/doi/abs/10.1080/00029890.1995.12004620 (accessed 16 August 2021).
32. Ryrie, S., Daintith, J., & Nelson, R., (1991). *The Penguin Dictionary of Mathematics. Applied Statistics,* [online], *40*(1), p.179. Available at: https://rss.onlinelibrary.wiley.com/doi/10.2307/2347919 (accessed 16 August 2021).
33. S., D., & Karush, W., (1963). *The Crescent Dictionary of Mathematics. Mathematics of Computation,* [online], *17*(84), p.478. Available at: https://www.jstor.org/stable/2004035?origin=crossref (accessed 16 August 2021).
34. Salvy, B., (2010). The Dynamic Dictionary of Mathematical Functions. *Lecture Notes in Computer Science,* [online], 13–13. Available at: https://link.springer.com/chapter/10.1007%2F978–3–642–14128–7_2 (accessed 16 August 2021).
35. Sharakshane, A., (2019). *Educational English-Russian Mathematical Dictionary*. [online], Available at: https://litgid.com/catalog/

angliyskiy_i_dr_yazyki/uchebnyy_anglo_russkiy_matematicheskiy_slovar/ (accessed 16 August 2021).

36. Soil Science, (1957). *A Short Dictionary of Mathematics.* [online], *84*(3), 266–267. Available at: https://journals.lww.com/soilsci/Citation/1957/09000/A_Short_Dictionary_of_Mathematics.20.aspx (accessed 16 August 2021).
37. Spiridonova, N., (2021). Pedagogical conditions for forming bilingual mathematical competence in basic school students. *Education & Pedagogy Journal*, [online], *1*(1), 75–86. Available at: https://edujournal.tspu.edu.ru/en/archive.html?year=2021&issue=1&article_id=8155 (accessed 16 August 2021).
38. Tanzy, M., Tanova, O., Kara-Sal, N., & Mongush, A., (2017). Electronic dictionary of mathematical terms as a means of successful mathematics training of children with disabilities in bilingual conditions. Bulletin of Krasnoyarsk State Pedagogical University named after V.P. *Astafiev,* [online], *42*(4), 12–21. Available at: http://www.kspu.ru/upload/documents/2017/12/19/84802632e043b0bda673ae0dbc23c3ea/tanzy-tanova-kara-sal-mongush17417.pdf (accessed 16 August 2021).
39. Turner, L., (2021). From the War Against Errors to Mathematics After the War: Public Discourses on a New Mathematical Dictionary. *Mathematical Communities in the Reconstruction After the Great War 1918–1928,* [online], 119–150. Available at: https://link.springer.com/chapter/10.1007%2F978–3–030–61683–0_5 (accessed 16 August 2021).

Index